Failure

Why Science Is So Successful

Stuart Firestein

OXFORD
UNIVERSITY PRESS

OXFORD
UNIVERSITY PRESS

Oxford University Press is a department of the University of
Oxford. It furthers the University's objective of excellence in research,
scholarship, and education by publishing worldwide.

Oxford   New York

Auckland   Cape Town   Dar es Salaam   Hong Kong   Karachi
Kuala Lumpur   Madrid   Melbourne   Mexico City   Nairobi
New Delhi   Shanghai   Taipei   Toronto

With offices in

Argentina   Austria   Brazil   Chile   Czech Republic   France   Greece
Guatemala   Hungary   Italy   Japan   Poland   Portugal   Singapore
South Korea   Switzerland   Thailand   Turkey   Ukraine   Vietnam

Oxford is a registered trademark of Oxford University Press
in the UK and certain other countries.

Published in the United States of America by
Oxford University Press
198 Madison Avenue, New York, NY 10016

Library of Congress Cataloging-in-Publication Data
Firestein, Stuart.
Failure: Why Science Is So Successful / Stuart Firestein.
pages cm
Includes bibliographical references and index.
ISBN 978–0–19–939010–6
1. Failure (Psychology).   2. Science—Philosophy.   3. Discoveries in science.
I. Title.
BF575.F14F567 2015
501—dc23
2015009156

1 3 5 7 9 8 6 4 2
Printed in the United States of America
on acid-free paper

Contents

Acknowledgments

So many people deserve to be acknowledged for their many contributions that I am terrified of failing to be inclusive. Among the most important, though, is Alex Chesler. Right from the beginning Alex has been involved, as a teaching assistant for my first course in Ignorance while a graduate student in my laboratory and continuing with conversations over the years about failure and ignorance and the role they play in science. Alex could easily have coauthored this book with me, and we discussed precisely that on numerous occasions. However, the rigors and responsibilities of starting a new laboratory (at NIH) and a new family (two children with wife Claire, who was also a graduate student in my laboratory) made it impractical. But his mind and imprint are all over this book.

I was incredibly fortunate to have the opportunity to spend a sabbatical year (10 months in all) as a Visiting Scholar in the Department of the History and Philosophy of Science at Cambridge University. This book would have been finished a year earlier if it had not been for that sabbatical. And it would have been much the worse for it. The colleagues I interacted with there, the lectures that I attended, the long conversations over beer in the pubs (yes, it really is like that) added so much to this book that I can't imagine how I thought I could have written it before going there. A year's sabbatical does not make one a philosopher or historian, but it does develop an appreciation for those pursuits and a sense of the value that they bring to understanding science and how we do it, and even why we do it.

Although the entire Department welcomed me in the most inclusive way, I have to single out Hasok Chang, who seems to have inherited the unenviable job of being my liaison in the Department. Generous is not nearly sufficient to describe his treatment of me. Not only with his time, but more importantly with his ideas and perspectives, and questions and critiques. You will see that I reference him several times in this book. He is one of the most important thinkers, and writers, and doers in science today—by which I mean the entire effort, from teaching to experimenting to contextualizing to chronicling and documenting it. I am thrilled that he and his brilliant and equally generous wife Gretchen continue to count me as a friend and colleague.

Many others at Cambridge listened to me rant about failure and responded with thoughtful and challenging ideas. Many read portions of the book here and there and commented freely. I have stolen mercilessly from them. I can hope only that they see it as the flattery which it is. I am also happy to be able to say that I was welcomed as an honorary Fellow at Kings College. There I shared meals and wine and conversation with people who have perfected the epicurean art of blending sociality with intellect. The opportunity to have dinner or lunch on any given day with scholars of music, Russian literature, mathematics, classics, biology, or psychology was for me like being the proverbial kid in a candy store. I cannot thank the Fellows at Kings College enough for so many of the memories of the most memorable year of my life.

The Alfred P. Sloan Foundation and the Solomon R. Guggenheim Foundation provided funds to support my time in Cambridge and for many of the other expenses associated with producing the manuscript. I am deeply indebted (not literally, fortunately) to them for their show of faith and their interest in the subject. I hope that they are proud of what has come from their investment.

As with *Ignorance* before it, I find it a bit ironic to thank people for their contributions to *Failure*. Believe me, though, they are responsible for whatever is successful here. Many colleagues read over versions of the manuscript from the early stages to late in its development. Among them I especially include Anne Sophie Barwich, a newly minted PhD who can be intensely critical and immensely fun at the same time, Charles Greer, Matthias Girel, Peter Mombaerts, Jonathon Weiner, and Matt Rogers, and Brian Earp, a student and now good friend I met while at Cambridge.

I am also quite fortunate in being a member of a writing group, called Neuwrite, composed of scientists, from graduate students to lab heads, and writers, from students to professionals, all of whom are interested in the unique problems of writing good science for a wide audience. Remarkably, this group has continued to meet regularly for over seven years, and from it have emerged books, magazine articles, essays, newspaper pieces, films, and short stories that have appeared in a wide variety of venues, both online and in print (see http://www.columbia.edu/cu/neuwrite/members.html). Several chapters of this book have been workshopped by this group, and I have gotten many useful and challenging comments from the group and many of its members individually.

I am very lucky to have the backing of Oxford University Press, in particular Joan Bossert, who has been a great supporter, exceptional editor, and very good friend and martini partner; and the excellent staff of publicists, editors, and production people who worked on *Ignorance* and now on *Failure* and must be wondering what in the world could be coming next.

I was incredibly fortunate to have leased a small house from the Duncans, former Cambridge faculty members, on the aptly named Eden St. This house provided the perfect setting for reading, thinking, and, most importantly, writing. I tended to a growing community of diverse birds who visited regularly in the backyard. It still makes me a little homesick not being there.

Finally, I must acknowledge, even though that word hardly captures the debt I owe them, my wife and daughter, who went through *Ignorance* with me and then followed right along through *Failure*. They read this manuscript many times, kept me from straying more often than you could imagine, and contributed many important ideas. Their unfailing faith and support in one mad project after another is of course well beyond priceless. Assuming they are not completely crazy, they are the very best two people in the world.

Failure

Introduction

Perhaps the history of errors of mankind, all things considered, is more
valuable and interesting than that of their discoveries. Truth is uniform
and narrow . . . but error is endlessly diversified.
—Benjamin Franklin

This book has failure written all over it.

Literally of course, but metaphorically as well. So failure will stalk this book, and it may occasionally win a round. But if I get it right, you'll understand that those failures are an important part of the book, an absolutely necessary ingredient. A book on failure cannot just be a lecture; it must be a sort of demonstration as well. And so now, by some sleight of hand, I have at least partly inoculated myself against failing by telling you that the theme of the book is how important failures are. Come to think of it, that's also a theme—that we must make and defend a space for noncatastrophic failure, a place where failure can happen regularly.

This book follows, and is kind of an extension, of another book I wrote recently called *Ignorance: How It Drives Science*. As you can see I'm carving out a nice, neat, little niche for myself. It might seem like I am becoming a merchant of despair. In fact I find both of these subjects to be uplifting. Although ignorance and failure are commonly thought of in a negative light, in science they are just the opposite: they are where all the interesting action is. This will be a key point in this book—that failure in science is fundamentally different from all the other failures you've read about in self-help and business books and articles in *Wired* and *Slate*. It is a kind of failure we don't appreciate enough. Not understanding this, not appreciating failure sufficiently, leads to distorted views of science and denies one a surprisingly useful but rarely considered version of failure. I hope that on this one point I don't fail you.

Science, the great intellectual achievement of modern Western culture, is often depicted as resting on pillars of great foundational strength and intellectual might.

These pillars are variously identified as KNOWLEDGE and REASON, or FACT and TRUTH, or EXPERIMENT and OBJECTIVITY. Quite impressive. Students are regularly asked to approach science in the reverential way that these ponderous pillars demand. Perhaps such pillars are the correct depiction for textbook science—the stuff that is frozen in time and that generations of these same poor students have been required to master, by which we usually mean *temporarily memorize*. But then there is current science, the real stuff that goes on every day in laboratories and in minds across the world. Science rests, I am afraid to say, on two somewhat less imposing sounding pillars—IGNORANCE and FAILURE.

Yes, that's it, the whole tremendous edifice. The costly research programs, the years of education, the dedication of cadres of PhDs, teetering on top of Ignorance and Failure. But without these two, the entire business would come to a standstill. In fact, Ignorance and Failure are not so much pillars as engines that propel science forward. They make it at once a reckless endeavor and a conservative process, a creative enterprise composed of mind numbing reams of data. I understand that this view of science, beholden to Ignorance and Failure, is probably not the common perception, and that few outside of practicing scientists will immediately recognize the truth of this proposition. But I bet that anyone who has made a career of science, reading this now, is nodding in agreement. Indeed, every scientist to whom I have mentioned that I am writing a book on failure has immediately offered to contribute a chapter! Remarkably, most of us make a pretty good living doing this kind of work and virtually every scientist I know loves their work. So how could that be, composed as it is mostly of ignorance and failure—with perhaps a dash of accident or serendipity thrown in?

It may seem that I'm putting you on here, pretending to reveal some dirty little secret just to grab your attention. But the thing is, it's not a secret at all: it's general knowledge, inside of science. Somehow outside of the scientific establishment it seems we do a very bad job of letting everyone else in on what we do. So many things just get taken for granted that it never occurs to us to make it explicit. You know more or less what lawyers do, what accountants do, what journalists do, what car mechanics do—even if you couldn't do any of those things yourself. But when I tell a crowd of my daughter's friend's parents that I'm a scientist, all they want to know is what I do. Actually *do*, during the day, every day.

One curious thing about this book was that it never coalesced into a linear argument with some internal logic driving it forward. I didn't start the chapters in any particular order, and I kept working on them in no particular order. They are more like essays than chapters, each a reflection on some aspect of failure and science. The famed immunologist and science writer Sir Peter Medawar wrote a piece for the *Saturday Review* titled "Is the Scientific Paper Fraudulent?" His claim was not that scientific papers were untrue, but that they were constructed in a way that did not reflect the actual experimental or intellectual processes at work. They were reconstructed in some narrative order intended to drive home the point but were not an accurate record of the way it actually happened. This book is sort of just the opposite. It has not been put together in some carefully logical order that builds to a convincing and unassailable argument. It's more a collection of ideas, some of which I hope are new to you. They were to me.

One of the things I do hope this book will accomplish is to show science as less of an edifice built on great and imponderable pillars and more as a quite normal human activity. I don't mean by knocking it down a peg or two, but rather by building it up as a remarkable and surprisingly accessible way of seeing the world. Science is accessible to everyone because really, at its core, it is all about ignorance and failure, and perhaps the occasional lucky accident. We can all appreciate that.

ONE

Failing to Define Failure

A real failure does not need an excuse. It is an end in itself.
—Gertrude Stein

I have chosen this deceptively simple sounding statement, so typical of Gertrude Stein, to open this book because it gets so quickly to the heart of the matter. It challenges, right from the beginning, our idea of what a failure may be. What kind of a failure is Stein talking about here? What makes a "real" failure? Are there "unreal" failures, or lesser failures?

Like so many important words, *failure* is much too simple for the class of things it represents. Failure comes in many flavors, and strengths, and contexts, and values, and innumerable other variables. Nothing just stands alone as a failure without knowing something more about it. In the famous *Encyclopédie* of the French enlightenment, Diderot and d'Alembert (1751–1772) under the entry for *erreur*, which seems intended to cover failure as well, caution that there is no way to develop a general description or classification because *erreur* comes in so many forms. I started this project with what I thought were a few clear ideas about failure and its value in the pursuit of scientific explanations. What surprised me was how quickly those few ideas generated dozens of questions.

There is a continuum of failure, not just one narrow kind. Yes, there are failures that are just mistakes or errors, and they may often be no more than an unfortunate waste of time. There are failures from which you learn simple lessons: be more careful, take more time, check your answers. There are failures that can be taken as much larger life lessons: a failed marriage, a failed business venture; painful but perhaps character building. There are failures that lead to unexpected and otherwise unavailable discoveries: they often seem like serendipity, an accidental failure that opened a door you didn't even know was there. There are failures that are informative: it doesn't work this way; there must be some other way. There are failures that lead to other failures that eventually lead to some kind of success about learning why the other paths were failures. There are failures that are good for a while and then not—in science you might think of alchemy, a failure that nonetheless provided the foundations of modern chemistry.

Failures can be minimal and easily dismissed; they can be catastrophic and harmful. There are failures that should be encouraged and others that should be discouraged.

The list could go on. But I don't want to get sidetracked into a lengthy polemic trying to define failure, which would surely fail. We'll come upon all sorts of failures as we proceed, and we would do best to think of them as discoveries, not contradictions. Rather, I want to focus on the *role* that failure, in all its many identities, plays in science and how it contributes to making it such a successful enterprise.

Stein seems to be complaining about the common response to a failure—which is apology. Failure as mistake, unintended or unavoidable or because of some shortcoming that you are responsible for. Failure as the result of stupidity and naiveté that requires excuses and apologies. Why did you let that fail? Can't you do any better than that? Or, perhaps less antagonistic but no less disappointing, failure as inevitable. Well, that wasn't likely going to work. What did you expect? What a stupid thing to have even tried. And so forth. Stein, in that first simple sentence, identifies all these bad failures, useless failure, failures that demean failure.

Instead, how about failure that stems not from ineptitude, inattention, or incapacity. (True, even those occasionally turn out to reveal something unexpected and sometimes wonderful. But I wouldn't depend on them. Sloppy indifference can get you only so far.) A real failure is different from all those that need or are accompanied by an excuse—because it needs no excuse.

So what are good failures? Ones that need no excuse and are an end in themselves? Not really an end in the typical sense—that is, not an end where you give up trying anything else. Rather an *end* in the sense of something new and valuable. Something to be proud of and therefore requiring no excuse, even if it was "wrong."

Are there really such failures? Of course there are the mistakes we learn from, the errors that can be corrected, the failures that can be turned to success. But I'd like to take a chance here and venture that Stein meant something deeper than that. That she really meant meaningful failure. In the limit, this could mean that you might produce nothing but meaningful failures for your entire life and still be counted a success. Or at least never need to apologize. Is that really possible? What are these magical failures?

I have two possible answers. The first is that failures that are ends in themselves are interesting. Interesting is another word that one has to be careful about. It's easy to use, but then it's kind of vague and subjective. Is there anything that's interesting to everybody? I doubt that. But if we take interesting as a descriptor rather than an identifier—that is, a quality of something and not necessarily a particular thing itself—then we can perhaps come to an understanding. When the same Gertrude Stein was asked to write a piece about the atom bomb (shortly after its use in WW2 and, as it turns out, shortly before her death in 1946), she responded that it held no interest for her. She liked detective stories and that sort of literature, but death rays and super weapons were not that interesting because they left nothing behind. Someone sets off a bomb or some weapon of mass destruction that kills everybody and ends everything. So what's to be interested in? Certainly better if it didn't happen, but if nothing is all you're left with, then who cares? So maybe it's what's left that could make something an interesting failure. Good failures, we could call them Stein Failures, are those that leave a wake of interesting stuff behind: ideas, questions, paradoxes, enigmas, contradictions—you know what I mean. So that's one kind of successful failure I'm pretty sure about.

Here's the second idea. Is it the actual failure that is the end in itself? Or is it the willingness to fail, the expectation of failure, the acceptance of failure, the desirability of failure? Can you imagine making failure desirable? Can you imagine aiming at failure? Can you appreciate making failure your goal?

You *can* if you have the right idea about the word failure—what I hope to convince you is the scientific version of failure. It is more than a stupid error, more than a shortcoming on your part, more than a miscalculation, more even than a chance to improve. Yes, more even than failures as life lessons. I know we all believe that a failure can be valuable if you learn something from it. After all, that's what we call experience. But how about a failure that does not aim at later self-improvement? How about failures that really are *an end in themselves*?

In this sense virtually all of science is a failure that is an end in itself. This is because scientific discoveries and facts are provisional. Science is constantly being revised. It may be successful for a time; it may remain successful even after it has been shown to be wrong in some essential way. That may seem strange, but good science is rarely completely wrong, just as it is never really completely right. The process is iterative. We scientists hop from failure to failure, happy with the interim results because they work so well and often are pretty close to the real thing.

Newton was famously wrong about two little things—time and space. They are not absolute. Gravity is not explained by the attraction between the centers of massive bodies, although it looks that way and can be usefully described that way. To the extent that we can explain it at all, it seems to be best understood, for now, as an emergent phenomenon of mass creating curvature in space. An imperfect but useful analogy is the way a heavy bowling ball on a mattress causes a depression and things placed on the mattress tend to fall toward it, as if they were being attracted to it. But Newton's failure in that one regard, even though it seems like a fundamental part of the theory of gravity, is not at all fatal to the success of his work. His equations quite accurately describe action at a distance between two bodies—sufficiently well to calculate how to dock a rocket with a space station orbiting some 250 miles away and moving at a speed of 17,000 miles per hour.

Nonetheless, there was a nagging inconsistency in Newton's model over what appeared as two different kinds of gravity. This inconsistency was what needled Einstein so much that he was ready to take a most unintuitive, illogical perspective. Although it's not exactly how Einstein thought about it, these two kinds of gravity are most easily experienced as the loss of gravity—weightlessness. One of them can be felt as distance from a massive body (the weightlessness experienced in outer space), and the other is due to acceleration (the weightless feeling you would have in a rapidly dropping elevator). They seem to be from two different and unrelated causes—the mass of a nearby body and the force resisted by inertia, or acceleration. Two hundred and fifty years later Einstein essentially corrected the failure of that part of Newtonian mechanics by showing that in the correct inertial frame, one that does not assume absolute time or space, the two kinds of gravity are the same.

Granted, it turned out to be a rather major correction, requiring a Copernican-sized shift in our point of view. But as with Copernicus it didn't require throwing everything else out. We continue to live our everyday lives in a Newtonian world where space and time seem sufficiently absolute, just as we continue to live most of our lives in a pre-Copernican world where the sun "rises" and "sets." That oversimplifies the story a great deal (see Notes), but the point is that Newton was successfully wrong and it was the very failed part of his model that led to Einstein's remarkable insights. Pretty good work.

A failure can be even less successful—that is, wholly incorrect—and still useful. An example from biology might be the longstanding principle known as "ontogeny recapitulates phylogeny." This tongue twister of a phrase, coined in 1866 by the "father of embryology," Ernst Haeckel, is simply a slightly bizarre attempt at making a complicated concept memorable by forming a jingle about it. It means that over the course of its development an embryo in the egg (or uterus) appears to go through all the stages of evolution of that organism. For example, mammals early in embryonic development have what appear to be gill-like structures, making them look a bit like fish. These structures eventually develop into our jaws and other muscle and bone groups of our heads and throats but have nothing directly to do with respiration, as gills do for fish. In fact, Haeckel's concept is completely wrong, even though it held sway for decades and led to many advances in embryology. Not only is it wrong about embryology, it is wrong about evolution. We didn't evolve from fishes (or apes for that matter); we shared a common ancestor that evolved into both of us, in the case of fish some 500 million years ago, and in the case of apes only about 85 to 90 million years ago.

Nonetheless, this failed ontogeny-phylogeny concept gave rise to important ideas about how development proceeds in clearly established stages, and that structures do evolve from earlier forms, possessing a common ancestry even if a contemporary divergence. Haeckel's work was painstaking and actually started the branch of science we today call embryology. In particular, he introduced comparative anatomy and development—that is, the notion that we can learn a great deal by making comparisons across species. This showed crucially that not only were species related but that their development proceeded in a similar way along certain principles. The value of this "failure" to modern biology cannot be overestimated. On the other hand, it remains damaging in that there are many people who still believe in it because they were taught it as schoolchildren. You remember, the silly business about having had a tail when you were an embryo.

You could object that Newtown's and Haeckel's failures eventually led to successes and were not therefore really ends in themselves. I think that's too much to ask of failure. Failures like these not only lead to greater insights, they often lead to very unpredictable insights. They force us to look at a problem differently because of the particular way in which they failed. This could be considered the case with Einstein's recognition that Newton's little failure was actually a fundamental misconception about time and space. We expect success to lead us to even greater success. What may not be so obvious is that failure can do the same.

These then are what I would call the failures that need no excuse, that stand shoulder to shoulder with success. They are the packing material, the innards, of science, and not giving them their full due is to miss more than half of what science is about and how it works. The big job I hope to do here is to remedy that.

. . .

There are many trivial things that can, and have, been said about failure. They are the kinds of aphorisms commonly found in Chinese restaurant fortune cookies. I'll sum them up in a paragraph and then we can get on to the interesting parts of it—the much wider and deeper functions of failure that are undeservedly ignored or, worse, thoughtlessly rejected as undesirable.

So then here we go: failing is part of succeeding. Failure builds character. Those who haven't failed haven't tried. You never know yourself until you've had a failure. You have to learn how to pick yourself up and get back in the game. And so on. I'm sure you can think of other, similar platitudes. And they're all okay advice, especially when you have someone on the phone who is really distraught over a recent failure in love or work or sport. Sure, failing is part of life and managing it is important for your happiness. And there are innumerable books loaded with mostly trivial advice about how to do all that. So let's us be done with it.

What we are interested in here, subtly but importantly different from those earlier instances, is where and when failure is actually an integral part of the process. Where it deserves to be right beside success, where it doesn't just make for an uplifting story of the young lad or lass who succeeds with perseverance, but where failure really has to be there for the process to occur properly. It is the difference between Edison (Thomas) failures and Einstein (Albert) failures. Edison claimed he never failed, just found 10,000 ways that didn't work. But eventually he succeeded. And of course it probably wasn't 10,000 wrong tries, but the actual number doesn't matter—it was a lot of trying and finally succeeding. This is good advice for an inventor, less so for a scientist. Einstein lived on failure, his own and those of others, not just ways that didn't work. His working failures were deep inconsistencies, failures of theory, failures that produced understanding even more than success. No failure, no science.

Now this is not true of all other grand human endeavors. You don't have to fail first in business to later become rich, you don't have to fail at writing to gain success as a novelist, and you don't have kill a few people to become a good doctor. In none of these endeavors is failure required—although it may happen, and unfortunately often does. Successful people may try to convince you that failing was a key to their success; they will espouse uplifting narratives to document it, and even write whole self-help books to aid you on your journey through failure. But it seems that way to them only retrospectively, because they failed and then they succeeded. The people who just succeeded right away don't have the kind of narrative that makes for good reading, and they rarely have any counsel you can use. It has been reported that when James Michener ("Tales of the South Pacific," 1947) was asked how to become a successful author, he replied, "Try to arrange to have your first novel turned into a musical by Rodgers and Hammerstein." Good advice.

Failure in all these endeavors is not uncommon, but it is not necessary. Not so in science. Failures are as informative as successes, sometimes more so, and, of course, sometimes less so. Failures may be disappointing at first, but successes that lead nowhere new are short-lived pleasures. *Conclusion* has a curious double meaning in science. We use it often as the heading of a section in our publications. We have the Methods and Results and, of course, the Conclusions (although "Conclusions" is now often called "Discussion," which seems humbler). In this context the word refers to what you can deduce or infer from the data—that is, what we have succeeded in finding out. But it also means *ending*, and we almost never mean, or want, that. Most of the time the "Conclusions" are themselves riddled with new questions. And many of those questions arise because some of the experiments didn't yield the expected results. They failed. Enrico Fermi, the pioneering nuclear physicist, would tell his students, "If your experiments succeed in proving the hypothesis, you have made a measurement; if they fail to prove the hypothesis, you have made a discovery."

In science you not only have to have the stomach for failure, you actually have to enjoy the taste of it.

If you accept the contention that failure is both an inevitable and a desirable part of science, then it is sensible to ask how much failure. After all, it can't be only failure, or at least I'm pretty sure it can't be only failure. But I think we commonly underestimate the amount of failure that is acceptable. Just to get some sense of the scale of failure let's look at it in other places and see what the ranges of tolerance are. We could begin with the natural world.

Nature's greatest predators—the kings of the jungle, the sea, the air; the killing machines of National Geographic specials—well, it turns out that they are successful on just about 7% of their attempted pursuits. You may think that a lion, killer whale, or red-tailed hawk can go out and bag some poor defenseless animal any time it gets a craving for a snack. In fact, 93% of the time they fail to capture their prey, which is why they have to be wily and nearly always on the hunt. Not only that, but they generally hunt around the edge of the herd, picking off the sick, the lame, and the old. There's a reason for that: the failure rate for bagging some nice young and still juicy creature is even higher. Nonetheless, we still think of them as being at the top of the food chain and anoint them as kings of their niches. It seems from the biological perspective you can put up with a lot of failure and still make a decent living. (I suppose you could turn this around and say that the *prey* are successful a remarkable 93% of the time, but that is a difficult calculus since a prey animal can fail only once. And scientists are hunters, not prey, I hope.)

Evolution itself is a marvel of failure. Well over 99% of the species that have ever made an appearance are now extinct. Species continue to go extinct, some scientists believe at a currently alarming rate. How, out of all this failure, could the marvelously complex creatures that we observe have emerged? Could all of these remarkable animals and plants and ecosystems have been created by failure? Perhaps it is not so hard to see the attraction of creation narratives and why they elicit widespread belief when the alternative explanation is failure. But, like it or not, that is how it works.

Darwin's great insight was that evolution proceeds by random changes in an organism's makeup, followed by a selection process that favors beneficial changes. It thereby wipes out over time the useless or harmful changes and even the status quo. Today we understand that these changes are due to mutations in genes and these mutations are overwhelmingly failures. The failures perish, most of them immediately, the less failed ones after hundreds of thousands or even millions of years. But eventually they fail. A billion or more years of evolution is primarily a record of failures.

And it doesn't end there. The actual mechanism for evolution—the nuts and bolts of it, what we glibly call random mutation—depends itself on failure. Sperm and egg cells copy DNA from the parents to the offspring. DNA is a molecule with a structure famously identified as a double helix, like two spiral staircases wrapped around each other. It's the double part that is the key to DNA's hereditary function. Each of the two helices is a copy of the other. If they split apart, as they do in egg and sperm cells, then each helix, using some enzymes and other chemicals inside the cell, can fashion a new partner helix for itself. That is, it can copy itself. But the copying process is not flawless; it makes mistakes. These mistakes are what we call random mutations. They are random because the chemical process of copying is simply imperfect; it doesn't favor any particular kind of mistake. Some of the mistakes result in changes to a gene, one section of the DNA molecule, that are fortuitously beneficial, and the resulting offspring with this improved gene has some advantage over others with the older model gene. Stronger, faster, smarter, whatever. But since the process is essentially a copying mistake, a failure, most of the time the change is harmful, or at best useless. Without this faulty copy mechanism there would be no evolution, nothing for natural selection to work on. From what can be seen only as an overwhelming tide of mistakes and failures, the living world emerged. All the complexity, all the apparently clockwork precision of life, from the developing embryo to the most elaborate ecosystem—all of it is due to failure at an almost unimaginable scale. When you have a few billion years to mess about you can put up with a lot of failure.

Perhaps a more pedestrian, but more immediate, illustration might be athletics. This is an activity where success seems to be important, and failure is to be avoided. So, how acceptable is failure in sports? In the game of baseball a position player's salary—as opposed to a pitcher's—is generally keyed to their batting average. This number is the percentage of times the player reaches base safely by hitting the ball. It is calculated by dividing the number of hits by the number of at-bats (the number of times the player came to bat and therefore the number of opportunities to make a hit). Because the baseball season goes on for so long and player careers often last a dozen or more seasons, this batting average can be calculated to three significant decimal places. Thus the famous Joe DiMaggio of the Yankees had a lifetime average of 0.325. The decimal point is typically dropped, and one says DiMaggio had a "325 lifetime batting average." Joe DiMaggio was one of the all-time greatest baseball players, and perhaps along with Ted Williams of the rival Boston Red Sox (344 lifetime average), they were the greatest hitters ever. But what their averages tell us is that nearly 7 out of 10 times that they came to bat . . . they failed. They struck out, grounded out, flied out, or were out in any of a number of ways, and they simply went back to the dugout and sat down until their next chance to bat.

(To be precise, there was also the possibility of a walk, which allows the batter to advance to a free base because the pitcher threw the ball out of the strike zone four times and the batter didn't swing. But in baseball statistics a walk doesn't count as a batting opportunity and so does not affect the batting average. In fact, Ted Williams walked a remarkable 2,021 times, while DiMaggio walked only 790 times. There are many complicated reasons behind these numbers, some of which have to do with player skills and some of which are more a matter of strategy and other subtle reasons that are not so relevant to this argument. Lest you worry that I am about to slip into a long polemic about some baseball minutiae, I will resist and stop here.)

DiMaggio had 6,821 plate appearances during his 13 seasons, and he hit safely in 2,214 of those. But he was out 4,607 times. Nearly twice the number of times he was safe. Even more impressively, Ted Williams, in 19 seasons, came up to bat 7,706 times, hit safely 2,654 times, but was out 5,052 times. The two greatest hitters in the history of baseball had a combined total of nearly 10,000 failures!

Only a few players hit regularly at 300 and above, and these are the highest paid players in the game, commanding salaries in excess of $10 million per year. Ten million a year to fail 7 out of 10 times, dependably. Clearly failure can be a good bet.

So what's the answer? How much failure is acceptable? Of course there's no number or precise quantity to be calculated. But we can see from just these few examples that the acceptable rate is likely much higher than you would have imagined. In the limit, success need occur only once; failures can occur again and again as long as your resources, or your life, doesn't run out. And even a dependable failure rate of 80–90% could be considered successful. Experience, that much-valued attribute of the learned, is after all the result of not getting it right the first time. Niels Bohr described an expert as "a person who has made all the mistakes that can be made in a very narrow field." Notice that it is not someone who has been a success in some narrow field.

TWO

Fail Better

Advice from Samuel Beckett

Ever tried. Ever failed. No matter. Try again. Fail again. Fail better.
—Samuel Beckett

I wrote this chapter after being reminded, by English novelist Marina Lewycka, of this quote from one of Samuel Beckett's lesser known, later short stories. Since starting it I have learned that the quote has become a staple of self-help and business books, headlined by one of the ubiquitous Timothy Ferriss manuals on how to be fabulous in no time at all with little or no effort. Then I found that, thanks to an article in *Slate* magazine, it has become the darling phrase of Silicon Valley and the so-called entrepreneurial set. My first thought was to accept having been scooped and jettison the chapter. But then I read the other pieces, mostly essays, out there that use this quote and realized that it was actually the perfect opportunity to illustrate how what virtually everyone else means by failure is different from what it means in science. And what better co-conspirator than Samuel Beckett.

The terse lines are generally taken to be a literary version of another of those platitudes on failure. The old "try, try again . . ." trope. But of course Beckett was rarely so simple. In one of my favorite literary descriptions, Brooks Atkinson in a *New York Times* review of *Waiting for Godot*, famously called the play "a mystery wrapped in an enigma." Not a bad overall description of Beckett.

Being unable to top that I will forgo, to your relief I'm sure, a critical interpretation of Beckett here. But there is something especially penetrating about that quote that is worth a few moments of our time to explore. Beckett offers an idea about failure that is not at all common, but that is very close to what I think is the scientific sense of the word.

The statement is typically succinct (12 words in 6 sentences!), but seemingly trivial. Perhaps an autobiographical life lesson about failing. It could be, except for its terseness, the first lines of a self-help book. Yes, I've tried, and yes, I've failed, but that will not stop me! I'll try again even if I fail again.

But then, suddenly there is that last two-word sentence. Fail better. Fail . . . better? Now what could that mean? How do you improve on failing? Is there a better way to fail? Is there a worse way to fail? Isn't failure just failure, and what's important is how you treat it, bounce back from it, overcome it? Beckett is trying again, not in order to succeed but to fail better.

Failing to write a popular novel—which he certainly had the ability to do, failing to repeat what had made him famous, failing just to try again without *trying to fail*; these options were not for Beckett. Failing better meant eschewing success when, or because, he already knew how to achieve it. Failing better meant leaving the circle of what he knows. Failing better meant discovering his ignorance, where his mysteries still reside. Try again, of course. But not to succeed. Try again, To Fail Better.

It is this unordinary meaning of failure that I suggest scientists should embrace. One must try to fail because it is the only strategy to avoid repeating the obvious. Failing better means looking beyond the obvious, beyond what you know and beyond what you know how to do. Failing better happens when we ask questions, when we doubt results, when we allow ourselves to be immersed in uncertainty.

Too often you fail until you succeed, and then you are expected to stop failing. Once you have succeeded you supposedly know something that helps you to avoid further failure. But that is not the way of science. Success can lead only to more failure. The success, when it comes, has to be tested rigorously and then it has to be considered for what it doesn't tell us, not just what it does tell us. It has to be used to get to the next stop in our ignorance—it has to be challenged until it fails, challenged *so that* it fails. This is a different kind of failure from that of business or even technology. There, it's "Make a mistake or two, sure (especially if it's on someone else's dime), because you can learn from those mistakes—but then that's enough of failure." Fail big and fail fast, the tech guys say. As if it were just something to get out of the way as quickly as possible. Movie executive Michael Eisner said in a 1996 speech, "Failing is good as long as it doesn't become a habit." Once successful, there should be no backsliding. But failure is not backsliding in science—it moves things forward as surely as success does. And it should never be done with. It should become a habit.

By trying to fail better, Beckett enlarges his sphere rather than shrinks it. It's nearly, but not exactly, the opposite of the process of trying to succeed, which is not necessarily succeeding, as trying to fail is not necessarily failing. Trying to succeed entails sharpening a technique, honing a strategy, narrowing in on the problem, focusing your attention on the solution. At times, of course, none of these is a bad thing. Indeed, in the day-to-day work of science this is the recipe for accomplishment—if by that you mean publishing papers and getting grants. There are many scientists who would say that is what science is about: that our job is to put pieces in a puzzle, and the more pieces you add, the more successful you are. It's hard to argue against this very pragmatic approach, which seems to be "successful" in the sense we discussed earlier.

Except to note that this process is driving science into a corner, separating it from the wider culture, failing to engage generations of students, turning it into a giant maw for facts and balkanizing the effort into smaller and smaller specialties, none of which have any idea what the others are about. We all recognize that there is something wrong with this. We can't keep up with the exponentially expanding literature of ever narrower details, we can't agree on what the right spending priorities are, we can't seem to affect public policy with our knowledge. We—scientists—are more and more a secret society of oddballs and geeks, tolerated because now and again some gadget or cure drops out of the otherwise impenetrable machinery we are supposed to be controlling. And as long as the rate at which that happens is sufficient to satisfy the tax-paying public, then they'll keep supporting "whatever it is you guys do." This process may be successful in some narrow sense of the word, but it is doomed to run out of steam, or at least to bore us all to death.

The alternative? Fail better. But how does one do this? Not easily, as Beckett reminds us. Try writing a grant proposal in which you promise to "fail better." Try getting a job with a research strategy that lays out your program for failing better. Try attracting students to come to your lab where you promise them every opportunity to fail better.

I know how crazy that sounds, but it is of course exactly the right way to proceed. If you are reviewing a grant, you should be interested in how it will fail—usefully or just by not succeeding. Not succeeding is not the same as failing. Not in science. Thomas Edison labeling his failures to perfect the light bulb as 10,000 ways of not succeeding is the right thinking for technology and invention. And it's not a bad mantra for the Silicon Valley folks, because at least it tells them to be patient and put up with not succeeding for a while. But it's not the same as failing better.

The right question to ask a candidate for a faculty position who has just presented his or her five-year research plan is, what is the percentage of this that is likely to fail? It should be more than half—way more than half, in my opinion. Because otherwise it is too pat, too simplistic, not adventurous enough—especially for a young scientist. And, really, a five-year plan that anyone should believe, especially the person presenting it? Who among us could predict anything five years into the future? What kind of science would science be if it could make reliable predictions about stuff five years out? Science is about what we don't know yet and how we're going to get to know it. And no one knows what that is. We often don't yet know what we don't know. And that deep ignorance, the unknown unknowns, will only be revealed by failures. Experiments that were meant to resolve this or that question and fail to do so show us that we needed a better question. So what I want to know from a young scientist is, how are you going to set up your failures?

Failure is not something to tolerate while focusing on the bright side. Failure is not a temporary condition. It must be embraced and worked on with all the diligence that one is accustomed to putting into succeeding. Failing can be done poorly or it can be done well. You can improve your failing! Fail better.

How do you do this? Of course it would be silly if I had a prescription for failing, any more than if I had a surefire prescription for succeeding, since the whole idea is that there is no single way. That said, I can make a few personal recommendations, just to think about. First, I recognize that failing better is not easy in the present culture. Perhaps for this moment in history the opportunities failure creates will be best realized as a personal choice, as a stratagem that you adopt to make decisions about what outlier to investigate, about which crazy project you'll keep going a little longer than might be advisable. It is a momentary kind of subterfuge, the secret drawer where you keep your unfundable, but not unloved, ideas. You know, that drawer that can be opened only by nudging out one side first and then the other, and back and forth. Paying attention to failures—not for the purpose of correcting them—but because of the interesting things they have to say, because they are humbling and make you go back and reconsider your long-held views. No failure is too small to be ignored or go unregarded.

A key breakthrough in the discovery of a family of enzymes known as G-proteins was made by eventually realizing that the dishwashing soap used on the experimental glassware was adding trace amounts of aluminum, and that this was a crucial cofactor in the G-protein's activation. No one would have suspected any such thing. It caused years of frustrating failures of many experiments, but it finally led to one the most important discoveries in pharmacology—and a Nobel Prize. This is only one of hundreds, if not thousands, of such stories, big and small, about productive failures that led to an otherwise unconsidered finding.

Of course the problem is that we have stories only about the failures that eventually led to a success. That's not because they're necessarily a better kind of failure but rather because those are the kinds of stories we tell, so that's the data we have. Yes, there are cases where a failure simply says, oops, wrong tree (to bark up); let's move on. And they are not without value. They can be just as elegant and creative and thoughtful as things that worked out. They deserve a place of honor—and we'll get to that issue shortly.

It may be harder to recognize the intrinsic value of failure when it eventually results in a success, in the sense of a positive finding, such as the identification of the G-protein. But there are two ways failures have an intrinsic value beyond the correction they provide. The first, and perhaps obvious one, is that there is no way to predict which way they will turn out. They may lead to a success, or they may hurtle down a cul-de-sac. Or, more often, they may lead to a partial success that will fail again a little further down the road, leading to another correction. This iterative process—weaving from failure to failure, each one being sufficiently better than the last—is how science often progresses.

Failures also don't just lead to a discovery by providing a correction (e.g., control for aluminum in glassware by using plastic); they lead to a fundamental change in the way we think about future experiments as well—and, in this case, the way we think about enzymes and how they work and how to discover them. So now we know that trace metals (the list, to date, includes copper, iron, magnesium, zinc, and others) in vanishingly small quantities are important in proper enzyme function. And they may come from unexpected places, like glassware. This failure then is data. That the experiments eventually worked because the aluminum was controlled for actually serves to confirm the failure. We don't think of confirming failures, but that's what we often do. Then we selectively remember the success, since it's such a relief, and the failure goes unsung.

It's not just young scientists who have become failure-averse, although it is most painful to see it happen there. As your career moves on and you have to obtain grant support you naturally highlight the successes and propose experiments that will continue this successful line of work with its high likelihood of producing results. The experiments in the drawer get trotted out less frequently and eventually the drawer just sticks shut. The lab becomes a kind of machine, a hopper—money in, papers out.

My hope of course is that things won't be this way for long. It wasn't this way in the past, and there is nothing at all about science and its proper pursuit that requires a high success rate or the likelihood of success, or the promise of any result. Indeed, in my view these things are an impediment to the best science, although I admit that they will get you along day to day. It seems to me we have simply switched the priorities. We have made the easy stuff—running experiments to fill in bits of the puzzle—the standard for judgment and relegated the creative, new ideas to that stuck drawer. But there is a cost to this. I mean a real monetary cost because it is wasteful to have everyone hunting in the same ever-shrinking territory. Yes, we have that old saw about looking under the lamppost because the light is better there, but now and again you have to venture out into the darkness, beyond the pool of light, where things are shadowy and the likelihood of failure is high. But that's the only way the circle of light can expand.

I attended a seminar on research trends in Alzheimer's disease where neurologist David Teplow of UCLA showed a graph of the numbers of papers published, beginning just before 2000, on something known as "Aβ protein." This was when a couple of labs published some work suggesting that Aβ (said as A-beta) was an important contributor to Alzheimer's disease. Indeed, they claimed that it was *the* causative factor. Within months, and continuing still, there has been an exponential increase in papers about Aβ. From a few citations a year, the Aβ protein now appears in over 5000 papers per year! It has turned out, in all likelihood, to be a chase after a phantom—if the idea was that ridding a patient of Aβ would cure Alzheimer's. It's as in the Chinese proverb where the first dog barks at something and a hundred bark at his sound. But this bandwagon effect, in which some finding gets published in a high-profile journal and everybody goes chasing after it, can be seen in virtually every field of science.

And it is a chase. Not a thoughtful exploration. Not an attempt to unravel a mystery. Not even the "new" and promising line of research it is advertised as being. Mostly it is a direction that is suddenly on the funding map and therefore one to pursue. By the way, to keep the record straight, Aβ is surely involved in Alzheimer's disease, but it is no longer considered the causative factor by most researchers and has even been found to serve a beneficial purpose in normal brains. Generally it is considered unlikely to be a good candidate as a drug or treatment target. Don't even ask what this all cost.

How will this change? It will happen when we cease, or at least reduce, our devotion to facts and collections of them, when we decide that science education is not a memorization marathon, when we—scientists and nonscientists—recognize that science is not a body of infallible work, of immutable laws and facts. When we once again recognize that science is a dynamic and difficult process and that most of what there is to know is still unknown. When we take the advice of Samuel Beckett and try to fail better.

How long will this take to change? I think it will require the kind of revolutionary change in our thinking about science comparable to what Thomas Kuhn famously, if perhaps a bit inaccurately, identified as a paradigm shift, a revolutionary change in perspective. However, it is my opinion that revolutionary changes often happen faster than "organic" changes. They may seem improbable or even impossible, but then once the first shot is fired, change occurs rapidly. I think of various human rights movements—from civil rights for black Americans to universal suffrage for women. Unthinkable at first, and then unfathomable as to what took so long. The sudden fall of the supposedly implacable Soviet Union is another example of how things can happen quickly when they involve a deep change in the way we think. I'm not sure what the trigger will be in the case of science, but I suspect it will have to do with education—an area that is ripe for a paradigmatic shift on many levels, and where the teaching of science and math is virtually a poster child for wrongheaded policies and practices.

Max Planck, when asked how often science changes and adopts new ideas, said, "with every funeral." And for better or worse, they happen pretty regularly.

THREE

The Scientific Basis of Failure

Things fall apart, it's scientific.
—David Byrne

Failure is supposed to happen. Science itself says so. No less than the Second Law of Thermodynamics demands it. You don't argue with the Second Law of Thermodynamics. I mean really, could they have found a scarier name for something than The Second Law of Thermodynamics? Who wants to tangle with something like that? I've always felt it had a kind of Old Testament ring to it that says, don't mess with this fruit. But being scared off by that would be a real shame, because wrapped up in that scary moniker are a couple of elegant and essential ideas. And they are very useful for understanding failure properly.

This frightening sounding *Law* is merely the formal explanation of the term *entropy*. Now that may be another somewhat obscure-sounding word, but in fact it's a pretty simple and very intuitive concept—except as it was presented to you in your high school physics textbook. Here's what entropy is, mostly. You know how your desk, room, house, office, or car is always a mess no matter how much you try to organize it? Well the reason for that is entropy—and the Second Law.

You see, there are a limited number of ways that your desk, room, house, or car can be organized and orderly. But there are an unlimited, possibly infinite, number of ways it can be messed up. Books, for example, may belong on the bookshelf, and that's the one way that they are organized. But they could be virtually anywhere else in your house, and all of those places would be included in the disorganized column. If you think like a Las Vegas casino owner—there are a limited number of ways to win, and a huge, maybe infinite, number of ways to lose—then which is more likely to happen more often and where does the house place its bet? Same with your desk: lots of ways to be a mess, and only a few ways to be neat. Clearly, then, it is far more likely that it will occupy one of the numerous messy states and far less likely to be in one of the much fewer neat states. Now here's the kicker. The same thing is true for the whole universe. And that's what the Second Law of Thermodynamics says, and entropy is a way to measure all that disorder. We could call it the sloppiness factor, but in this case *entropy* sounds more elegant, even if a bit more esoteric. So the next time someone suggests that you clean up your desk, or car, or whatever, just tell them it's a hopeless battle against entropy and it's not your fault.

The same entropy factor is at work in failure. Failure is the expected outcome according to the Second Law of Thermodynamics. There are many more ways to fail than to succeed. Success, by definition, should be very limited. Failure is the default. Success requires an unusual, but possible, confluence of events in which entropy is temporarily reversed. This is reminiscent of Tolstoy's famous remark that all happy families are alike, but every unhappy family is unhappy its own way. You might call this the Anna Karenina factor, a kind of literary equivalent to entropy. Notice that Tolstoy felt the most interesting things to write about, to investigate, to explore, were the varied ways that families did not succeed, were not happy. His inspiration sprung from the myriad ways things went wrong. It's not so different in science.

Now because of all that variation, it is true that most of the failures will not be very useful ones. Again, that's just probability. But as in the case of a neat desk, by inputting some energy (e.g., cleaning it up) it is possible to at least temporarily reverse the Second Law here and there—locally, as the physicists would say. That's why, for example, we have this very organized state of matter called human beings, where a lot of energy goes into fighting the forces of disorder. Of course the Second Law always wins in the end and thus we get old and become physiologically more disorganized, also known as sick, and then terminally disorganized. Dust to dust.

Failures that are useful are a sort of little cheat on entropy. Not easy to do as you might imagine, cheating on the Second Law ... We do it through selection and intelligence and what might be known as feedback—or error correction in slightly more technical, but still I think understandable, terms. When you add up both sides of the ledger and include everything, then the overall entropy, the disorder, in the universe is still increasing in accordance with the Second Law. You are robbing Peter to pay Paul. But what the hell, that's how things work. Let's see just how.

Failures provide a certain kind of feedback that is then used in a process we call error correction. With this simple loop in place, knowing that something doesn't work can be as valuable as knowing that it does. Of course once again there are probably many more ways that something won't work, so it's harder to design an experiment that will report narrowly on what isn't working. Harder it may be, but also probably more valuable. Often it takes a series of experiments to narrow in on the failure. The first ones leave many options open, from technical glitches to fundamental misperceptions. So you go back and try to think of all the things that could have led to the failure and then you try correcting for them. Often you won't even be able to determine what led to the first failure until you do more experiments that fail. You will have to do experiments just to reveal the true nature of the failure.

If all this sounds hopeless or exhausting, let me assure you, it's just the opposite. Tracking down failure is when you are most creative. Designing experiments to identify failure requires cleverness and cunning. You must be in the most critical state of mind. It is the closest a scientist comes to Sherlock Holmes. Nothing can be dismissed in the face of failure. The smallest clue could be the key element. What's missing is just as important as what's there. There is a universe of possibilities. And indeed, there are innumerable cases of important discoveries being made because the failed experiment revealed a new set of possibilities that you hadn't even realized were there. This is sometimes mistaken for serendipity, a notion that, since it's come up, I would like to take a moment to dispute.

Serendipity is a popular idea in science narratives. An absurd number of Nobel Prize winners claim, with modesty false or honest, that their discoveries were almost entirely serendipitous. But I think this is essentially wrong. A charming concept, the term *serendipity* was coined by Horace Walpole, a "man of letters," around 1754. He took it from a fairy tale called "The Three Princes of Serendip," in which three princes of what is now likely to be Sri Lanka or Ceylon, travel around more or less aimlessly and wonderful things happen to them. Even Walpole called it a silly tale, but serendipity, describing unplanned good luck, has recently achieved considerable popularity. Reading science reports in the newspaper, you might think that fully half of the reported discoveries were serendipitous. Sometimes it may be just gracious humility, but many times I think a scientist truly believes that there was a pivotal and even magical dash of luck that brought the discovery to him or her rather than someone else just as talented.

That may be, but it is critical to remember that, unlike Walpole's version of the fortunate and frivolous three princes, in science you have to be at work for these blessings to alight. Lawyers and financiers don't make serendipitous scientific discoveries; hard-working scientists do. And even there, it is rare for a scientist to make a serendipitous discovery in a field other than his or her own. In fact, most so-called serendipitous discoveries are made through failure. Something doesn't work the way you thought it should and exploring the reasons for that leads to the initially unexpected and now surprising result. It is the very intensity of tracking down a failure that forces you to reconsider what you're doing at the very basic levels. And the more you fail, the more you have to dig down to the basics, sometimes giving up cherished ideas and concepts that you were sure were established beyond any reasonable doubt. And then, bingo: there is the new answer, hiding behind all those failures. It may seem like serendipity, and you may feel like a charmed prince (or princess), but that's because you have clamored to that place where the unexpected replaces the expected. Okay, maybe it is a bit charming after all.

One of the classic examples of this sort of serendipity—that is, serendipity from failure—was the discovery of the cosmic microwave background radiation (CMBR). I say classic because it's a story of a yearlong failure that resulted in a Nobel Prize. To make it short, Arno Penzias and Robert Wilson, two astronomers working at the famed Bell Labs in New Jersey in the 1960s, built a new super-sensitive radio telescope to record faint signals from far-off regions of the galaxy. But the instrument was plagued with static noise—of just the sort you get when your radio is not perfectly tuned. They considered all sorts of sources for the noise "artifact"—nearby New York City, nuclear bombs, weather, pigeon crap on the outdoor parts of the apparatus (described by Penzias as "white dielectric material")—but none of it could account for the persistent, if faint, noise. After a year or so of this, they were fortuitously put in touch with a theoretician at Princeton named Robert Dicke, who had predicted that this very sort of noise would be expected as the energetic remnant of the Big Bang that began the universe. In fact, the discovery of this background radiation essentially proved the Big Bang theory.

Importantly, others had made similar predictions as early as the late 1940s, and a group of Russian cosmologists had a similar result at about the same time as Penzias and Wilson. You see, it was all ready to happen; it just needed a good failure to move it along. Penzias and Wilson were awarded a Nobel Prize in 1978 (but not Dicke; that's another story) for "discovering" the microwave background radiation. Now the story is often told as if it were just so much dumb luck. But in fact many scientists worked intensely over a period of several years to figure this out, and a huge amount of science had been done to make the answer sensible when it finally appeared.

As Louis Pasteur (himself a recipient of considerable "serendipity") noted, "Chance favors the prepared mind." I would add a corollary to that, blurted out over a dinner of Indian food in London by my colleague Tristram Wyatt: "Failure favors the prepared mind!" We were, of course, deep in an animated discussion of the finer points of scientific discoveries when this serendipitous thought just appeared.

So if you think serendipity is an important factor in scientific discoveries, you really should be thinking that it is failure that is the important ingredient. Science progresses not because of simple and charming serendipity, but because of bruising accidents and crashing failures and a lot of tough repair work.

Failure, from this perspective, is a challenge, almost a sport in the way it gets your adrenaline going. Figuring out why this or that experiment failed becomes a mission. It's you against the forces of failure. You will need stamina, strategy, skill. There is great urgency, but you must apply tremendous patience. Can you see the elevated state of mind a good failure can put you in? Can you see how the possibility of an important discovery is more likely to happen in this state than when you are simply tabulating the results of a "successful" experimental run? Indeed, failure truly favors the prepared mind, and it prepares that mind.

I don't want you to think, as you well might, that having written a book titled *Failure*—with the theme that failure is good and too often ignored, to our detriment—that I am in any way against success. As they say, you can't argue with success. All I'm saying is that now and then a good argument is just what's needed, and if you can't do that with success, you certainly can with failure. Second Law be damned.

FOUR

The Unreasonable Success of Failure

The most exciting phrase to hear in science is not "Eureka," but, "Hmmm, that's funny . . ."
—Isaac Asimov

In a book on failure, one thing we have to admit is that science has been preposterously successful. Particularly over the last 14 or so generations. It has not gone unnoticed by philosophers, historians, journalists, or even scientists. Physicist Eugene Wigner famously delivered a lecture called "The Unreasonable Effectiveness of Mathematics in the Natural Sciences" (and published it as a paper in 1960). Therein he marvels at how successful mathematical formulations have been at describing the physical world—often in ways that were not intended or expected by their original authors. Galileo's mathematically expressed laws of falling bodies in Northern Italy could be extended through the calculus of Newton to planetary objects in space and even to stars and distant galaxies, indeed to concentrations of mass anywhere in the universe. It is not at all clear why this should be so, and in the last paragraph Wigner suggests that it "is a wonderful gift which we neither understand nor deserve. We should be grateful for it and hope that it will remain valid in future research and that it will extend, for better or for worse, to our pleasure, even though perhaps also to our bafflement, to wide branches of learning."

His philosophical ruminations on this subject have sparked many commentaries and the production of like-minded articles from scientists in many branches of study besides physics—from mathematics to biology to computation. Physicist David Deutsch, in his large and comprehensive book on modern science, *The Beginning of Infinity*, points to the rapid acceleration in the rate of scientific and technical advances in the last 400 years, especially compared to the rate of progress in the first 5000 years of human history, when humans had all the same brain power we currently have. Imagine, the Bronze Age lasted some 2000 years. For 2000 years, more than 50 generations of people—people with the same brain you have—were born, lived, and died without an appreciable change in technology. I don't have to tell you how many times I've changed phones/computers/cars and so on in just the last 10 years.

Philosophers have taken very seriously the question of why science is so successful, some going so far as to claim that there should be a scientific explanation for why it is the case. Philosopher J. R. Brown, in a paper titled "Explaining the Success of Science," cites the ways that science is an overachiever, including its technological accomplishments, its handiness for building bridges and curing diseases, its entertainment value (so many good discovery stories), and its success at extracting tax dollars from us all (Michael Faraday, asked what electricity might be used for, said he didn't know, but that surely the queen would figure out a way to tax it—which was of course prophetic). More seriously, Brown lists three generally accepted ideas about successful theories in science: (1) they are able to organize and unify a large variety of observed phenomena; (2) they improve the understanding of existing data over previous theories; and (3) they make a significant number of predictions that pan out—that is, they are better than guessing. These seem like reasonable conditions.

It is important to say here, though, that the word *truth* has not made an appearance. I mean, isn't science about finding out the truth? Isn't it successful because it discovers the way things truly work? When you get right down to it, isn't that what we mean by a success in science? That is, we found something out that was true. Actually it is more like TRUTH in capitals. Were it only so easy.

In science we often try to figure out if some thing or event causes another thing or event, usually because we have noticed that the first thing is often followed by the second thing. The important question that has to be asked is whether the first thing is both *sufficient* and *necessary* to cause the second thing. Sufficient *and* necessary. Often, quite often, you can show only one or the other. Nature is inexplicably stingy when it comes to causes versus correlations. Correlations are the weak stepchildren of causes—and they play tricks on you all the time. Because two things happen close to each other in time does not necessarily mean that one is a cause of the other, or for that matter that they owe anything at all to each other. On the other hand, sometimes this association maybe a real clue to causation. This is where necessity and sufficiency come in.

Thus you may find that event A is sufficient to cause event B, but it is not necessary—other things can also cause B without A. Or A is necessary, but not sufficient—A has to be there, but by itself cannot cause B to happen. I bring all this up because we should ask this question about truth and success if we think one explains the other. Is truth both necessary and sufficient for success in science?

If you give this a little thought—well, maybe more than a little—you will find that, remarkably, it is neither necessary nor sufficient. A surprising answer, at least to me. Science has many successes to its credit that later turned out not to be true, or not entirely true. And quite often we are happy to have something that is insufficient to explain everything about an observation, but will work for the time being. I would even go so far as to say that big truths, packed with both necessity and sufficiency, could be an impediment to science. At least the search for such things would be an impediment to the dynamic nature of science practice. Ultimately we may demand the kind of truth that comes from showing both necessity and sufficiency, but who knows when "ultimately" will arrive?

It is generally accepted that the greatest success in biology is Darwin's explanation of how species originated and thereby how living things came to be what they are—how evolution worked to produce "from so simple a beginning endless forms most beautiful and most wonderful ..." But Darwin did not have not the last word. Evolution, the concept, has continued to evolve itself, incorporating new findings in fields from paleontology to molecular biology to computer science and others. Darwin's theory is under constant revision. Unquestionably he got the basics right, the incredible insight that from randomness and probability, with a feedback process (not of course what he called it), order of the highest level can emerge. But much of what we call evolutionary theory today was missing in Darwin's writings. Darwin was a careful thinker and cautious to a fault. It took him more than 20 years to publish, and it is entirely possible he would never have gotten to it if not forced, ironically by competition, to "rush" his manuscript into publication. So there isn't much that is wrong in *The Origin*, but it is certainly incomplete.

Perhaps most glaringly, there is no mention of the word *gene*, a term that was then unknown. Darwin admits his ignorance as to what the hereditary "particle" might be, and his later speculations were all quite far off the mark. It's not that the information wasn't there. More or less contemporaneously with Darwin, Gregor Mendel demonstrated how generational inheritance works through crossbreeding, transferring genes for particular traits between individual plants. But Darwin had no less information about inheritance than Mendel, didn't lack some special technology that Mendel had developed, was not short on cash or resources. There is no reason Darwin could not have embarked on an experimental program similar to Mendel's. In fact Mendel's work is often highlighted for its simplicity (although that is sometimes overstated as his classic experiments were very labor intensive and required more than 7 years of work). And Darwin was famous for his botanical interests and abilities and he also experimented with plants. So how come he didn't figure out genes? And it wasn't only Darwin. The entire community of proto geneticists failed to recognize Mendel's work as being about anything more than plant hybridization. Its brilliant insight into the laws of inheritance went unrecognized for some 35 years—until its so-called rediscovery in 1900. This is not a knock on Darwin. He has had the company of many great minds throughout history in this surprisingly endur-ing mystery of not seeing the obvious solution right under your nose.

I use Darwin as an example because of his vaulted posi-tion in the pantheon of biologists and scientists generally. Among revolutionary thinkers Darwin ranks with Galileo, Newton, and Einstein. And his confederates on that list, or any list of accomplished scientists you care to make, were no more immune to failure. There are serious failures—whether errors, false ideas, missing insights—in all of their work. But it is precisely from these flawed enterprises that we get the unreasonable success of science. If this sounds like a contra-diction, it should. There are many truths in contradictions.

Of course it's easy to see all this in hindsight. But we should be wary of that phrase. While Darwin may be in our hindsight, we will one day be in our students' hindsight. It's hard to know what we are missing right under our noses that will seem so obvious to a generation hence. This is where the connection between ignorance and failure is made. Our failures tell us about our remaining ignorance, and our igno-rance produces our failures. And so on, in this cyclic engine that occasionally spits out knowledge of the first order.

Will science continue to be so successful? Is this accel-eration of the last 400 years just the beginning, as David Deutsch would have it? Is it sustainable? Is acceleration built into science, discoveries leading to ever more discoveries in an exponential piling on—the more we know the more we can know? Will science, begun in the 1600s and still dedicated to its foundational methods, survive the political buffeting of the modern world? Science has appeared before, only to disappear for long periods of time. Something that we would recognize as science emerged at one time in Asia, Arabia, Mesopotamia, Egypt, Rome, Mayan Mesoamerica—and then for mostly obscure reasons just stopped. Just like that. And it could happen again, here and now. Science may look like it is here to stay, too big and too successful to fail, but consider an example as recent as the 70-year regime of the Soviet Union: much of Soviet science was distorted beyond recognition in the supposed service of the people. If the Soviet view had prevailed, science would be quite different from what we experience today. And only a few years before that, Hitler's Germany dismantled what could arguably have been the greatest scientific establishment the world had ever seen, simply by expelling half of their scientists for being Jewish, and forcing the rest to work on producing destructive technologies.

To be sure this doesn't happen again, we should look carefully at how it has happened in the past. A short historical perspective will be instructive here.

From the 8th to the 12th centuries CE, while Europe muddled through the perhaps overdramatically named Dark Ages, science on planet Earth could be found almost exclusively in the Islamic world. This science was not exactly like our science today, but it was surely antecedent to it and was nonetheless an activity aimed at knowing about the world. Ruling caliphs bestowed on scientific institutions tremendous resources, such as libraries, observatories, and hospitals. Great schools in all the cities covering the Arabic Near East and Northern Africa (and even into Spain) trained generations of scholars. Almost every word in the modern scientific lexicon that begins with the prefix "al" owes its origins to Islamic science—algorithm, alchemy, alcohol, alkali, algebra. And then, just over 400 years after it started, it ground to an apparent halt, and it would be a few hundred years, give or take, before what we would today unmistakably recognize as science appeared in Europe—with Galileo, Kepler, and, a bit later, Newton.

So what happened? This is a question hotly debated among historians of science. From a very Western comparative perspective, many have proposed that it was simply that Arabic science went as far as it could and it took the Europeans to pick up the torch and carry it onward to the pinnacles of knowledge it has reached today. Of course this is just so much ideological chauvinism. European mental life of the 12th century was hardly a paragon of free thinking and uninhibited inquiry.

Patricia Fara, a Cambridge historian of science, taking a very even-handed perspective, points out that Islamic science had a different purpose, and therefore a different approach from the scientific view that developed in the West. Islamic science was based on accumulating knowledge for the purpose of understanding God through his universe. It was, like many early traditions that preceded modern science, interested in the well-being of the soul and understanding the godliness of the universe, and much less in manipulating it. Great libraries were built and stocked with encyclopedias that were studied by generation after generation of students. Knowledge was the path to God, and gaining it was the process of salvation. The purpose of science was to accumulate, classify, and organize the knowledge that one would need to be spiritually fulfilled.

Fara points out that the culminating event in Arab science was the publication of the *Book of Healing* by Abū ʿAlī al-Ḥusayn ibn ʿAbd Allāh ibn Al-Hasan ibn Ali ibn Sīnā, or Ibn Sīnā for short. He was a Persian polymath who also went by the Latinized name Avicenna (a sort of phonetic bastardization of Ibn Sīnā). The *Book of Healing* was not a medical text, but rather an encyclopedia of all that was known. Reading it would "cure" your ignorance. Noble as the task might have seemed to Avicenna, compilations of facts do not propel science. Indeed, they have a way of stifling it. And the last thing you want "cured" is your ignorance. Islamic science didn't falter; it reached its goal.

Now this is not trivial because there has been, and still is, a great deal of importance attached to possessing the "complete worldview." It is mostly the province of religion, but science is too often used in its service as well. As you may have imagined, I consider this idea a bankrupt strategy and most likely the result of an emotional weakness built into some less useful portion of our poor hunter-gatherer brains. It is not what makes science work, but what causes it to stall. Once again we find that the best intentions—to collect the facts, to vet great, or at least prolific, thinkers, and to establish truths—is antithetical to the practice of science. Science grows in the mulch of puzzlement, bewilderment, skepticism, and experiment. Any other way leads to the end, to ossification and unfounded beliefs.

"Know these things and you are saved" may have been the purpose of early protoscientific endeavors, but modern science does not promise salvation it cannot deliver. I don't mean to claim that we don't owe a tremendous debt to the Greek philosopher-scientists who developed geometry and the earliest versions of astronomy and navigation, or to the Arabs who developed algebra and preserved and organized the great writings of the ancients, or to the Chinese who apparently knew about magnetism long before the West, or to the clergy and scribes of the Middle Ages who translated the Arabic texts into Latin thereby preserving a continuity of thought from the ancient world to the Renaissance. But in the end their models of science failed to embrace ignorance and failure as their driving forces, and the effort was guided not by experiment and empiricism but by philosophical and spiritual desires.

I'll end this chapter with a letter written by Albert Einstein in answer to a query about why he thought science developed so much more in the West than in the East. At this time Einstein was famous; he had become virtually the symbol of science during one of the greatest historical periods of scientific advances. But still he seems keenly aware of the fragility of it all. We should take his warning seriously.

April 23, 1953

Mr. J. E. Switzer,

San Mateo, California

Dear Sir,

Development of Western Science is based on two great achievements, the invention of the formal logical system (in Euclidean geometry) by the Greek philosophers, and the discovery of the possibility to find out causal relationship by systematic experiment (Renaissance). In my opinion one has not to be astonished that the Chinese sages have not made these steps. The astonishing thing is that these discoveries were made at all.

Sincerely yours,

A. Einstein

(Reprinted in Derek J. de Solla Price, *Science Since Babylon*, New Haven, CT: Yale University Press, 1961)

FIVE

The Integrity of Failure

The purpose of science is to not fool yourself—and you are the easiest
person to fool.
—Richard Feynman

There is yet another, and perhaps not so obvious, way in which failure is key to the scientific enterprise. This has to do with the integrity of science. And the integrity of scientists. I don't mean scientific misconduct or fraud, which I am sure is what you think of first. These are important concerns, although I have to say that the reaction to the few notorious cases where some conflict of interest or fraud has actually been proven has resulted, not surprisingly, in an unwieldy web of policies that threaten to slow research to a crawl under the weight of innumerable administrative checks. But that rant must wait for another day, because it is not this kind of legalistic integrity that I want to talk about here.

Put most simply, how reliable is success if there is not sufficient possibility of failure? Success becomes more successful, and often more interesting, the harder it is to obtain, the more likely the process that led to it could have led instead to failure. I always think of golf in this regard and a brilliant sendup of the sport by the late Robin Williams. In it he tries to explain the rules to someone, and each rule makes the game seem more difficult and improbable than the last. It starts simply enough—you use a stick to hit a ball into a hole. In a series of more and more absurd rule clarifications he tells us: No, no, a tiny little ball, into a tiny little hole with a really thin crooked stick—from 300 yards away. And somehow that's what makes it such a fun game to pursue.

In the same way, one could imagine that the whole institution of science, the entire infrastructure—the methodology, pedagogy, daily practice, and literature—all exist to make failure possible, even probable, without it being absolutely fatal. The risk of failure is not reduced, nor is the potential risk to career and standing. But if done right, failure can be an acceptable outcome. But what does that mean, *if done right?*

Andrew Lyne is an astronomer who devised a method for detecting planets around other stars and then thought he had found the first planet outside our solar system. On the eve of reporting this discovery at the American Astronomical Society, he realized that he had made a fatal error in his calculations and that he had not actually discovered the first exoplanet. He delivered his talk the next day, admitting his error, and was greeted by his colleagues with a standing ovation for his courage and honesty. His method, by the way, was correct, and it enabled the subsequent discovery of several exoplanets—by other researchers. That's failing right.

The possibility, even likelihood, of failure demands of the scientist a level of integrity and personal responsibility, a willingness to follow the data no matter how it works out, to take the result where it will go—including nowhere. And it means that at least part of the reason for choosing a certain question will be for its difficulty, its likelihood for failure. If this is not a crucial part of how the scientist proceeds, then the claims that a scientist, any scientist, makes about something being true are ultimately hollow, or worse, boring.

Another way to think about this is to consider what we expect from a scientist. The popular conception seems to be that scientists solve problems. They get answers to questions, and the better they are at that, the better a scientist they are. At the risk of being repetitive, you know of course that I don't think this is true at all. *Finding problems*—good problems, relevant problems, important problems—that's what a good scientist does. And where do these problems come from? Our failures, of course. Failures are the most reliable source of new and better problems. Why *better*? Because they have now been refined by the failure. We know a bit more about what we don't know. And with this new, distilled ignorance we can see more clearly what the crucial question is.

Is every scientist up to this standard? And note that I am not talking about a standard of ethics but rather a standard of courage. Perhaps not. But the majority must be for the whole system to work. Although my reasoning here may admittedly be a bit circular, science seems to be working, so I'm betting that the majority of scientists do meet this standard. Courage is not typically thought of as part of the scientific toolkit, but making a risky prediction, showing integrity in the face of mounting evidence against you, requires courage.

How do scientists come to this view? Not consciously, I think. And not, unfortunately, from the mostly superficial ethics courses that graduate students are required to sit through and which typically are reminiscent of driving school for traffic offenders—a rehash of virtually all the things you already knew or should have the common sense to know. Rather, this comes from observing and being involved in the practice of science. A practice where it soon becomes apparent that failure is the most common outcome. Even in routine lab work, funny things happen. Reactions that should turn blue, turn green. Hmmm. They should either have done nothing or turned blue. Now what? Graduate students and postdocs can be heard groaning all around you as they look at some newly developed photograph or computer output—that doesn't show what they expected.

Perhaps the best place to learn about this is the lab meeting. Usually a weekly affair that often involves circular baked goods with a hole in the middle and lots of coffee, the lab meeting is when everyone gets together and looks over each other's data. There are many formats. Some labs rotate around the members, each week another person presenting a prepared report, typically a PowerPoint presentation, the data they have collected over the last couple of months. In other cases, it is less formal and anyone with new data presents a result or two in a very rough form. Still other labs have everyone say something about their findings over the previous week—even if it was nothing but failures. I like the last one the best, of course. But whatever format is used, the lab meeting is mostly about failure, and it is the place where a young scientist learns how to manage failure, learns the value of failure, and actually sees it at work. Here the student will see a varied display of attitudes toward failure, from sweeping it under the rug ("Well, let's just keep trying until we get that to work") to confronting it and opening the discussion widely to possible solutions or interpretations. The latter immediately sounds like the better strategy—but there are times when going back and trying some more, not giving up too quickly, is also correct. So it's not black and white.

A friend and former postdoctoral fellow of the brilliant and pioneering Cambridge neuroscientist Alan Hodgkin once told me that the best way to get Hodgkin's attention was for things not to work. He would come into the lab each morning and pass by each person's workstation and ask how things were going. If the results were coming in more or less as expected, he would nod approval and move on. But if you were stuck, if the experiments were just not working, if the data were uninterpretable, then Hodgkin would take off his jacket, fill his pipe, and sit down for a long discussion. Of course what interested him was why it was not working. If it worked, it was as expected. Fine, but then let's move on. If it didn't work, then it was critical to find out if the problem was a trivial technical issue or, even better, a deeper misunderstanding that could reveal some deeper understanding.

It is in the daily practice of science, in the hum of the lab, in the interaction with a mentor (a cowardly one who whips the results mercilessly to fit the prized theory, or the braver head of lab who sees the failures as opportunities)—in all these ways the apprentice learns the integrity and courage of science.

Failure is also a test of dedication. It is a way to measure what you are passionate about and how deep that passion runs and how dependable it is. Science may seem methodical but it demands passion. Persistence in the face of failure is of course important, but it is not the same thing as dedication or passion. Persistence is a discipline that you learn; devotion is a dedication that you can't ignore. Persistence may overcome failure, but failure tests devotion. Science is not the bloodless undertaking that is often portrayed in the media, and it is when the failure level is high that you see the passionate demands it makes on its adherents.

These lessons will be clear to the serious student. And the results, at least in the long term, will also be clear. Because that is ultimately how science works. Reeling from failure to failure in order to make the successes more meaningful. If science is to produce something more than trivial knowledge it must be hard, it must be susceptible to what the late philosopher John Haugeland called the *collision* between the theoretical and the empirical—what we thought to be the case and what the experiments indicate is in fact the case. Unless this collision is possible, nay likely, then how much can science claim to have discovered by finding that narrow path that avoids the collision? T. H. Huxley famously quipped that there is nothing so tragic in science as the slaying of a beautiful theory by an ugly fact (Arthur Conan Doyle later put these same words into the mouth of the fictional Sherlock Holmes). And it happens a lot. But the flip side of that is there is nothing so inspiring as success snatched from the claws of failure. That is the success most worth celebrating.

SIX

Teaching Failure

The things I remember best from college were the questions I got wrong on the exams.
—Kathryn Yatrakis, Dean of Academic Affairs, Columbia College

What if we removed failure from the scientific equation? I have claimed that it is an absolute requirement. Can I support that contention? What would happen if we got rid of failure? Now we can't actually do that, but we can—and do—come close in two areas. One is education. (The other is funding, which I'll deal with separately). We teach only the successful science, not the failures. Is our science education working? This is pretty easy—no.

Here is a kind of case history described to me over lunch by a leading philosopher of science. It is a true story about his daughter who many years earlier, in the 8th grade, arrived home and announced that she wanted nothing more to do with science. Since she had been a good science student up to that moment, he sat down with her and asked why this sudden conversion. Well . . . they, she and her classmates, had been given a problem in a physics class having to do with pendulums. Given the equation that described the motion of a pendulum, she was asked to calculate its energy at the highest point of its arc and at its trough before swinging up in the other direction. After thinking about it for a bit, she was confused by her realization that at its highest point the pendulum bob was actually perfectly stationary and therefore would seem to have no energy at all. A not-unreasonable observation, and one that puzzled her. When she engaged the teacher and described the problem, instead of getting any kind of explanation, she was told to just go ahead and solve the equation by plugging in the correct number, and don't make a fuss over the details. "It will work out," she was told.

In fact pendulums have been an object of scientific interest since Galileo, and besides Galileo, there was Kepler, Liebniz, Newton, Huygens, Euler, and a host of unknown but clever clockmakers, who grappled with the mechanics of the pendulum for two full centuries before coming up with the final version of the equation that had just been thrown at her by her teacher. It is clearly not a trivial problem that any fool should understand. Unless of course you consider Kepler, Liebniz, et al. fools. Just because we now have the equation in hand does not mean that the process of understanding the phenomenon is transparent. It is the way to appreciate the difference between the kinetic energy of the swinging pendulum and the stored or potential energy it gains at the top of its arc, which is then converted back into kinetic energy on the downward swing. This fundamental insight that makes it possible to keep an energy balance sheet was, and is, crucial for understanding the cardinal principles of the mechanics that run the universe. The *two-century-long* record of failures in coming to the correct equation describing the swing of a pendulum lends as much, quite probably more, to understanding physics as does plugging numbers into an equation.

In case you think this is all a trivial argument over an obscure and entirely academic puzzle, the pendulum swing was the only way we kept accurate time until the invention of quartz movement in the 1970s (another 300 or so years later!). And Foucault's pendulum, now in the Pantheon in Paris, was the first demonstration that the Earth rotates on its axis and the sun remains fixed, contrary to the way it looks (earlier "proofs" were inferences made from astronomical observations). Even now, modern investigations of harmonic oscillators, important in dynamics and atomic theory, can be traced to the pendulum.

The point of this story is that removing failure from science education removes explanation and leads to pedagogically criminal behavior like "Just fill in the equation and solve it for full credit." A "teachable moment" was lost, and, indeed, had her father not cared so much, a child's interest in science could have been squandered for a lifetime. Every fact in science was hard won and has a trail of failures behind it. These failures shouldn't be hidden; they should be featured. First, because it is how science actually works. And second, because if students realize that if the likes of Newton, Liebniz, et. al. could have failed to grasp something, it might just occur to them that, "Hey, maybe I'm not so dumb after all."

Ernst Mayr, in his deeply perceptive book *The Growth of Biological Thought*, defends the importance of a historical perspective in science: "Only by going over the hard way by which these concepts were worked out—by learning all the earlier wrong assumptions that had to be refuted one by one, in other words by learning all past mistakes—can one hope to acquire a really thorough and sound understanding. In science one learns not only by one's mistakes but by the *history of the mistakes of others*" (italics added).

Failing to include failure is what a culture of testing begets. In a test, you don't ask for the 10 wrong answers that preceded, necessarily, the right answer. But those 10 wrong answers were a matter of reasoning. We talk about teaching critical thinking to our students, but then we give credit for memorized answers, not thinking. Critical thinking develops when you understand why people thought the wrong thing for a long time and came to the correct answer only by slow increment or sudden insight. And, really, I should say the *currently* correct answer, because there is almost surely a better one to come. Even in investigating pendulums, which now contributes insights to the study of chaotic activity. (There is an entire project called the International Pendulum Project [IPP], which publishes a 500-plus-page book divided into four sections covering scientific, historical, philosophical, and educational perspectives of the pendulum!)

Okay, so it's a disaster. We're busily alienating 13-year-olds and doing a first rate job of destroying any interest they may have in science. Surely no one intends for this to happen. Science teachers I've known tell me how they are miserable over drilling facts into kids' heads to prepare them for the next exam. They hate what I, and others, have come to call the bulimic model of science education—cramming gobs of facts into their heads so that they can puke them back up on an exam, then move on to the next unit with no measurable gain. They recognize that this is not teaching science, but they feel stuck with the system.

It would be easy to blame the system, and of course there is blame to be had there. But if you're going to assign fault, you have some obligation to suggest solutions. Science, more than any other subject, forces us to confront these problems. Because science does, after all, have a lot of facts. For all my ranting against the overly hallowed attitude toward facts, in this book and my earlier one, *Ignorance*, the truth is that science has facts, lots of them. There are some things you have to know. There is a part of science education that has to come to grips with the tremendous collection of factual knowledge that has accumulated over the last 400 years, much of it over the last 50 of those 400 years. The knowledge base is indeed vast and growing, and we have to bring students up to speed in a relatively short time. This is a great challenge. First because you have to make choices about what the critical things to know are, and then often you can't just teach a list of stuff that has no context or background. So then you have to teach yet some other things so that the really important things can be understood. Where do you start and where does it end?

This dilemma reminds me of the comedian Don Novello, who took on the fictional character of an Italian priest named Father Guido Sarducci producing fake news reports as a Vatican correspondent. In one skit in the 1970s Sarducci/Novello announces that he is opening the Five Minute University where, for $20, you can learn in five minutes what the average college graduate remembers five years after graduating—and get a diploma. It even includes a 30-second spring break. This brings to mind the brilliant scene near the end of *The Wizard of Oz* when the unmasked wizard is nonetheless grandly making good on his promises to Dorothy and her crew. You will remember that the scarecrow wanted a brain to think great thoughts. The wizard assures him that he has a brain, as good as any other, but the only thing he doesn't have is . . . a diploma.

Invoking some obscure authority, the wizard confers upon him a degree (the wonderful ThD, Doctor of Thinkology) whereupon the scarecrow rattles off an impressive string of facts and equations, proving that he can "think great thoughts."

These are satires, but they strike at the heart of the matter. Is being educated a matter of knowing a certain amount of stuff? And, if so, how much and what should it consist of? I suspect most of us believe that this is not what educated means, or at least we hope it means more than that, but it is nonetheless what we continue to provide to our children. And more importantly, and perniciously, how we judge them—and their teachers. Education is what's left over after you have forgotten what you learned in school. Many people have said that, in one form or another, since the beginning of schools.

I don't have the answer. If I did I would have told you long before this. But I do think there are places to start, and this is all we need to do right now. We have to experiment with the curriculum—and there will of course be failures. At least if we are doing it honestly and seriously, there will be. We don't have to do radical things. There can be incremental experiments just as there are in science. We may not need a "paradigm shift." It may turn out that a little adjusting would do the trick. We should fix what we can right away, and then go to work on the recalcitrant stuff that's still left.

What's easy? First, we could remove what is called the tyranny of coverage. This is the notion that a course in chemistry must cover as much of chemistry as possible in the couple of semesters or terms given over to it. The same for physics or biology or environmental sciences or whatever. Of course covering everything is an obviously absurd pursuit. We couldn't cover an entire field even in a virtually unlimited time. No scientist in any of those disciplines knows all there is to know in his or her own domain, let alone mastering three or four fields. Yet we use massive textbooks that attempt to comprehensively, if thinly, cover an entire field. And what will be remembered of this whirlwind tour through the edifices of physics, chemistry, or biology? Any of us who has had a high school (even AP) or college physics course can answer that question easily. Virtually nothing. I'm a professional biologist and I remember almost nothing from those physics courses. I know some physics because I need it in my work, but I've had to relearn what I found I needed. A lawyer presumably never thinks of it again.

Would we consider it acceptable to teach a course in literature with nothing but the Cliffs Notes? Of course not. But isn't that essentially what science textbooks are? Would it be okay to take a course on Shakespeare and read only the plot summaries? Can you imagine teaching a course on Joyce's *Ulysses* without including reading the original text in depth, bringing in background materials to place it in historical context, then considering the forward-reaching effects on literature and the novel? But those are precisely the elements we leave out in our science courses.

So step number one could be giving up *coverage* and just changing the balance a bit between fact accumulation and understanding through context. How about putting a narrative component into every unit? Nothing in science came from nowhere. Everything we know has a history attached to it, usually a rich history. Not even so much a *his*tory as just a story. A story of how a puzzle came to be posed and then answered and then a new puzzle arose from the answer. How about a story of the failures that led to the current best, if still wrong, answer that we have?

Here's an example. My friend, and I like to flatter myself by calling him my colleague, Hasok Chang in the Department of History and Philosophy of Science at Cambridge University, wrote a fascinating book on the invention of temperature. Among other things, it's about how science works by iteration, getting closer to a truth by making mistakes that are each a little less of a mistake than the one before it. Now, temperature would seem to be the simplest thing in the world to understand. We ask students to do science experiments that range from measuring and tracking the day's temperature to understanding heat transfer in thermodynamic engines. But do we ever ask them to consider how they would invent the thermometer they're using to do these experiments? It's not so trivial because there is a kind of circularity to it. How would you know what temperature a thermometer was measuring if you didn't already know some temperature that you could use as a standard? It's fine to discover that mercury or air in a tube expands with temperature, but is that a linear process where each millimeter of expansion corresponds to some number of degrees of temperature? Or is it more complex, where, say, colder temperatures expand the mercury proportionally less than warmer ones? Do air and mercury, or any other substance, react to temperature in the same way? How would you know if you didn't have at least one fixed point? And really, two would be much better.

Boiling or freezing points of water, you say? Not so simple in reality. We all know "a watched pot never boils," and that turns out to be truer than you might think. At least you can't get people—observers, if you prefer—to agree on exactly when a pot of water begins to boil. And even if you did, it would be at a different temperature depending on your altitude—in Denver water boils at a lower temperature than in San Francisco. Is that at all obvious? Would you even think such a thing if you didn't already have a thermometer to make that discovery? How come that's so, and what does that tell you about heat? For that matter, is it obvious what heat is? There have been many models of heat, most of them failures. Heat seems to flow, like a liquid, from one object to another—but always from a warmer object to a cooler one. Never in the other direction; that's how cold flows. Wait a minute—cold can't flow, cold isn't any "thing," just the absence of heat, right? Who decided on that? In fact, for a long time it wasn't clear if a thermometer was measuring heat or cold, and in many of the earliest thermometers the scale is upside down—by our current standards. Of course, it doesn't actually matter because there is no up or down in temperature—now there's a mind-bending thought for adolescent students.

Why is Fahrenheit a better scale than Celsius for living creatures? This is my own personal rant now, but I like it because it seems to drive almost everyone else in the world crazy. Fahrenheit is a biological temperature scale—it uses as its zero point the temperature at which blood freezes, and 100 degrees F is the approximate mean body temperature of mammals. So with Fahrenheit there are 100 equal divisions in the temperature range where most of us live our lives. Celsius, on the other hand, has a zero point set at water's freezing point and 100 degrees at its boiling point. This means that the Celsius scale has fewer gradations in the temperature range of living things, resulting in weather reports that are in half degrees and very often minus numbers. And then body temperature happens to be 37 degrees, an unwieldy and pesky prime number that can't be evenly divided when calculating what temperature you want for some reaction.

Celsius is a great scale for engineers and physicists, but for biology give me good old Fahrenheit.

As I said, this is my own rant, but you can also see that it could start a useful discussion about what temperature means and the fact that it is arbitrary but still describes a real physical condition. This is a fabulous model of how science works—from Newtonian physics to Darwinian speciation to Mendeleev's periodic table to Einstein's relativity and so on . . . Science can take something arbitrary and build a whole description of physical reality on it. It's a little bit like magic—which everyone knows is more fun and more engaging than just the facts.

Let's look at a really difficult problem in science education: equations and math.

Stephen Hawking, in his introduction to *A Brief History of Time*, says that his publisher told him that for every equation he put in the book, sales would be halved. He went for one equation anyway. By the way, this shows precisely the kind of faulty mathematical reasoning that is a typical product of our current math education program, because each succeeding equation would actually have a smaller and smaller effect on the actual number of books sold: half of a half of a half. So after the first equation you can include others at a much reduced cost, while still approaching zero sales, which perhaps was a subtle point that Hawking was making by including only the most expensive one.

I have been given similar advice in teaching, especially when it comes to biology students who are supposed to be the most math challenged of science students. Equations tend to lose the students, I am told. I disagree. Equations are valuable, not merely as a way to solve problems by plugging numbers into the variables, but because they are explanations written in a special shorthand. Like learning to type, it is worth learning something about that shorthand and how to manipulate it, and, more importantly, how to extract meaning and understanding from it.

Equations have stories to tell, stories of conflict and struggle, of failure and triumph, each elucidating some feature of the universe that was wrenched out by logic and thinking and experiment. And if you tell those stories, then they are not opaque or boring or ... formulaic. They aren't handed down from on high, and shouldn't be handed down to students as if they were.

Here's an example. I teach students something called the Nernst equation, named after its originator, Walter Nernst, a famous biophysicist of the late 19th century. Nernst was interested in batteries and electrical current and the fact that you could use salts to carry electrical charges. He figured out how to describe this mathematically so that for any given concentration of ions—sodium and potassium and chloride, most commonly—you could determine the voltage that would be created by distributing them unequally.

Nernst worked this out as a way to build some of the earliest batteries—and to some extent, standard alkaline batteries still work along these principles. But it wasn't long before physiologists realized that they could use the same equation to describe how electrical activity arises in the body. This is why we have instruments such as the electrocardiogram or the electroencephalogram or the electromyogram to measure electrical activity in heart, brain, and muscle, respectively.

So would it have occurred to you that an equation developed by a physical chemist could be used to describe how the brain works? Lord Kelvin, another physicist, used Nernst's equation in part to figure out how strong an electrical signal you would have to inject into the end of a wire in New York for enough of the signal to make it to London, traversing the Atlantic Ocean in the first transatlantic cable. Those same calculations also tell us how a signal gets from your brain to your big toe through a nerve fiber soaked in salty bodily solutions—not so different from the transatlantic cable in seawater. And we measure that signal today to diagnose diseases such as amyotrophic lateral sclerosis (ALS) or multiple sclerosis (MS). I could go on, but I hope the value in making these connections is obvious. How many microseconds after I said the words "Nernst equation" would your mind wander off to something, anything, more interesting? But if I started by telling you I was going to explain two types of batteries—the ones you use to power your gadgets and the ones that power the brain that uses those gadgets to entertain itself—all in one simple formula . . . well, that's got to be a bit more interesting, if only because it seems so unlikely.

Of course, there are some things you should know if you are going to be scientifically literate—much as there are some things you should know if you are going to be literately literate, or artistically literate. People in the humanities argue about this all the time. What are the books and writings that make up the canon, the basic things every literate person should have read or heard about? What are the iconic works of art and architecture? These arguments are often heated—so heated, in fact, that the rest of the world looks on and thinks, really, what are you people getting so worked up about? It's just books. This one or that one, what's the difference, finally? In fact, it's a great thing that we live in a society where we feel we can afford to have some people with enough expertise to have arguments nearly to the death about what the canon should consist of, what the great ideas are. This is not just a luxury, because if we leave these decisions to politicians or school boards—even well-meaning politicians and school boards—then we might just as well have a Department of Right Thinking. We need experts and they have to have different opinions, reasoned and informed opinions, about how to write a prescription for literacy. Here's a case where unanimity is not a good thing at all.

I believe we should have the same "discussion" in science. We don't have a right set of things to know such that if you know this and that you can be considered scientifically literate. I think there are such things, but we won't agree on what they are. That's because, like culture, it is fluid and changing. It has to be revised, constantly. It is an argument we should keep on having. It is an argument we should enjoy having. It is an opportunity to think carefully about what the absolute fundamentals are, about what we can know in common and for the common good.

The humanities have been having this argument for decades (centuries?). The sciences should pick up on this good idea and get a good argument started. And I don't just mean silly ones about evolution versus intelligent design, which are politically and culturally motivated, and full of disingenuous claims. I mean serious conversations among experts in astronomy, biology, chemistry, computer science, ecology, math, and physics about what are the real essentials to know. Kind of like that game about what you would take along if you had to be alone on a deserted island for a year. Richard Feynman was once asked what fact he would want to have survive a holocaust that would wipe out virtually all of humankind (these were the apocalyptic post-atomic bomb days). Without hesitation he chose the idea of the atom. His reasoning was that losing this one idea would put us too far back to recover in any reasonable period of time and with any reasonable likelihood of success. But having that one concept would allow us to rebuild all of physics and chemistry—and then biology. Here are some others I'd include (in alphabetical order):

Calculus
Cells
Chemical bonds
Entropy
Evolution
Fields
Genes
Inertia
Kinetic theory of heat
Periodic table

Someone else will have another list, and I might change this list. But that's the point. There is no official list, regardless of what the curriculum czars would like you to believe. The unchanging curriculum based on maximal coverage is a great business model for textbook writers and publishers. But it has very little to do with optimal methods of teaching science. Once completed, these textbooks offer a comprehensive treatment that needs only minor updating here and there over many years—in some cases, decades. Have a look at the copyright page of some textbooks. Their first edition can go back more than a decade. The changes are often just cosmetic and have the primary purpose of undermining the used book market. (You see, if they add or subtract a few bits here and there, then that changes the page numbers. This forces everyone to buy the new textbook since the teacher's syllabus will be keyed to those page numbers.) I make this charge, not to vilify textbook publishers, but to underscore the lack of values associated with the textbook strategy of communicating science to new generations.

I have until now been talking about science education in general, mostly focused on the primary school years between, say, ages 12 and 18. In the early grades all the kids, the boys and the girls, like science. But beginning somewhere around the 7th or 8th grade we seem to begin winnowing out most of the students until, by the 11th or 12th grade, fewer than 5% of them want anything to do with science ever again, let alone consider it as a career option. We have developed an extraordinarily effective system for disengaging the interest of the maximum number of students. I doubt this is what we want. Nonetheless, it is what we have.

But this isn't even new. As long ago as 1957, famed anthropologist Margaret Mead and her collaborator Rhoda Metraux published an article titled "Image of the Scientist Among High School Students" in the journal *Science*. It was the result of a study undertaken by the American Association for the Advancement of Science (AAAS) to determine then-current attitudes about science and scientists among America's youth. The space race was on, and there was considerable concern about whether there would be enough young people planning on a career in science to meet the nation's technical and competitive needs. Sound familiar? So will the results.

Aside from the quaint, but now jarring, gender biases throughout the paper (all the more astonishing for having been authored by two women), the results are not different from what you might expect to obtain in a class of today's high school students. Among the several questions in the survey, boys were asked to complete the following statement in a short paragraph: "If I were going to be a scientist, I would like to be the kind of scientist who . . ." Girls were given the option of completing: "If I were going to marry a scientist, I should like to marry the kind of scientist who . . ." Since most of the boys were interested in getting married, or at least getting a girl, the girl's responses were considered of equal or greater importance than the boys in determining the actual career decision! By the way, this reasoning predates slightly the rediscovery of the crucial role of female choice in evolution by Robert Trivers and others in the 1960s and 1970s, although it had been previously alluded to by Darwin (1871), R. A. Fischer (1915), and others.

Based on the study results, students envisioned the "official" image of the scientist as someone who is essential to our national life and the world; someone who is dedicated and brilliant; someone who works without concern for money or fame and can discover cures or provide technical progress and defensive protection. All in all, someone we should be grateful to. On the other hand, when asked about career (or marital) choices, these very same qualities had strong negative connotations. The students referred to careers that were boring, only about dead stuff (unless it was adventurous like space travel), overly dedicated to something outside the home and family and normal relations, had an "abnormal" relationship to money (i.e., not making any), and were overall too exacting and too demanding. Now it really sounds familiar.

Mead and Metraux go on to fault the teaching of science for the development of these attitudes. They bemoan the fact that science is taught without "any sense of the delights of intellectual activity." The scientist is portrayed as someone working for years in a dour mood who "shouts for joy [only] upon finally finding something out." Students work with dead plants, dead animals, and even more lifeless textbooks full of the ideas of long-dead men. Mead and Metraux recommend changes to the teaching of science—and hopefully the attitudes toward scientists—that include, among other things, teaching a more realistic narrative of scientific discoveries that is not so hero-oriented. Obviously I couldn't agree more. But these recommendations were made in 1957. Nothing has changed. If anything, the situation has worsened because now we extend this bankrupt approach to the university years.

Today almost every college and university has some sort of science requirement for all students as part of their dedication to a liberal, well-rounded education—before you get started on the serious business of your finance major. These courses for nonscience majors are likely the last science class, lecture, or book that more than 80% of the college-educated population will encounter for the rest of their lives. And what are they being left with? More of the same bulimic drivel that they had in high school. Worse, these courses are often shunted off to overworked and underpaid adjunct professors. It's not that these instructors are not dedicated and hard-working, but leaving it to the temporary teaching force exposes the careless attitude most science faculties have toward these courses.

Now the students who want to be scientists will go on to graduate school. And they will get a completely different view of science. I have made a list of some of the things I learned in graduate school about science.

- Questions are more important than facts.
- Answers or facts are temporary; data, hypotheses (models) are provisional.
- Failure happens . . . a lot.
- Patience is a requirement; there is no substitute for time.

- Occasionally you get lucky—hopefully you recognize it.
- Things don't happen in the linear or narrative way that you read about in papers or textbooks.
- The smooth "Arc of Discovery" is a myth; science stumbles along.
- If there is free food, get there early.

I had never heard about any one of these critical verities as an undergraduate—even in my advanced science courses. It seems then there is a certain part of science, of the scientific process, that is available only to trained elite scientists. But there is nothing on that list that is beyond your comprehension. I could dazzle, or more likely bore, you with innumerable facts about the brain or the olfactory system, things that you would find incomprehensible, even assuming you were interested. But none of these sorts of things made the list of critical things I learned as a graduate student. Everything on my list is completely accessible. But you don't know anything about it—unless you're also a scientist.

This then brings us to the question of why we should teach science to nonscience students at all. If we aren't going to teach them the things they would need to know to be scientists, then what are we giving them that is so special, so unique, so critical to their intellectual development? Why do we feel teaching science is a necessity? What do we hope to accomplish?

Science has for the past four or so centuries provided more and better explanations about nature than anything in previous recorded history. Mostly it has developed a strategy for finding stuff out and knowing whether to believe it or not. This is not the Scientific Method, but rather a large body of accumulated procedures and modes of thinking, as well as established facts, that give rise to a kind of intuitive but informed worldview not steeped in magic or mysticism. Not that there aren't certain governing principles, but even they are not inviolable. And mostly it is a mechanism for making mistakes that are productive and not catastrophic. It is in this body of mistakes, failures, temporary explanations, crazy ideas, theories, and all the rest that we are to find the richness of scientific thought.

What should we do? Teach the failed ideas of caloric, ether, and prequantum theories of atoms; evolution by design; phlogiston; vitalism; phrenology; and so on in our science courses? You know the reflexive response of many science curriculum czars will be that these ideas all turned out to be wrong, so why would we teach them? That may be, but an awful lot of very intelligent and brilliant scientists believed in them at one time or another—so they couldn't be that outlandish. Physicist Michael Faraday was surely an intelligent design believer; Lavoisier and many others took caloric theory seriously for a very long time; subtle forms of vitalism can still be found lurking here and there in biology.

The point is that these concepts did not become failed ideas overnight. They were legitimate proposals at one time—and not propounded by stupid, uninformed people. How did we come to realize that they were fundamentally incorrect, even while they were "good" explanations? How can you discriminate them from a legitimate and correct scientific idea? What about these theories is still true, and how could that be when we think they are wrong? What is the difference between right and true, or even more between wrong and true? How do new theories replace older ones, even though they may not be completely correct and are certainly never end points?

I know it sounds crazy at first, but aren't these "failures" in many ways the best case histories? They require students to engage ideas that were taken seriously by some of the most brilliant minds in science. And then turned out to be mostly wrong. They demonstrate how easy it is to be wrong, to commit to a failed theory or idea. They show brilliant scientific minds on the wrong track, and from the vantage point of history you can see why they thought these were reasonable ideas. Science progresses by groping and bumbling and occasionally finding something out that usually leads to more groping and bumbling, but of a better sort. That's the process, and it works. Remarkably well. Why not teach it that way, instead of removing the failures and just covering the boring aftermath (i.e., the facts)?

And if you feel the need to be less historical and more modern, then there are plenty of recent ideas that have gone sour and needed revision. It's not like science has a history of failure that's all in its past. We do it every day in labs all over the world. We do it because science is about what we don't know, and there remains plenty of that mysterious stuff to generate ever new, and better, failures.

This kind of thinking is what science education can impart to students. Science is always a hard-fought battle; it is marked by successes and failures, by boldness and timidity, by joy and sorrow, by conviction and doubt, by pleasure and misery. It is too much of a grand human adventure to be represented by manicured and dry, preserved textbook accounts. And, like all human adventures, it is riddled with very fine failures.

The Arc of Failure

That isn't even wrong.
—Wolfgang Pauli, in reference to a paper he thought worthless

The Great Achievements in Science are often recounted as an "Arc of Discovery."

It has that historical sweep we so enjoy, with heroic players and occasional intuitive leaps and flashes of brilliance, well-timed accidental discoveries, culminating in a final radiant advance that magically synthesizes decades, or even centuries, of painstaking work into a brilliant and now solid understanding of how things really are. It's mostly about big discoveries—the atom, the chemical bond, the gene, the cell, the transistor; or conceptual leaps like inertia, gravity, evolution, algorithms. Behind every fact that we teach in our science classes there is the notion of an arc of discovery that has carried us triumphantly forward to some profound

concept—evolution or relativity, quantum physics or genomics. Newton claiming, perhaps sarcastically, to have stood
on the shoulders of giants; Watson and Crick announcing
they had discovered the secret of life in the Eagle Pub, now a
Cambridge landmark. Science as heroic narrative.

Except that's almost never the way it is. There are two
things wrong with this story. First, it's generally not true.
Second, it propagates a heroic version of science in which all
the major advances have sprung from the genius of a few
individuals. Galileo, Kepler, Newton, Faraday, Maxwell,
Kelvin, and Einstein form a famous arc in physics. It traces
the birth and growth of physics, from inertia to mass to gravity to energy fields and thermodynamics, and it includes a
final overarching explanation of them all in space-time by
showing mass and energy to be equivalent.

Discovery arcs like this one have a soaring property that
suggests smooth, accelerating progression toward a fixed goal.
I don't have to tell you that this bears little resemblance to
the actual history. It is a distillation, a frankly fictional shorthand, a superficial summary at best, of a process that is far
more complicated—and far more interesting. It is a process
that is actually replete with wrong turns, cul-de-sacs, and cir-
cularities in which facts are declared right, then wrong, and
then sometimes right again. There are blank spots that represent long periods of no progress, if not total inactivity—also
called being stumped. So the arc of discovery narrative has

not simply left out the so-called minor discoveries along the way by hitting just the high points, although that is true enough. More importantly, it has left out the failures and the struggles that were every bit as much a part of the process as the big discoveries that made the list. For example, it has left out the ether theory that dominated thinking for more than a century. It has left out caloric theory, the heat substance that doesn't exist but played a critical role in getting us on the path that eventually became thermodynamics.

Perhaps even worse than what is left out, this arc mythology suggests that we have reached an end when we have not—there still is no unification between Newton-Einstein physics and that of the quantum subatomic universe (a second whole arc of discovery story in physics). And did you notice how guilty I just was, implying there is an arc of discovery we could label as Newton-Einstein physics—as if there were a direct, unbroken line through 250 years of historical time. It's very hard not to partake of this glibness.

Derek de Solla Price, a prodigy in the history of science who died at the untimely age of 61 and whose works are almost entirely, and wrongly, out of print, has pointed out that the history of science is unlike any other history as its entire successful past exists in its present as well. "Boyle's Law is alive today as the Battle of Waterloo is not." But it is only the *successful* past that is represented. And this has the unintended and not entirely positive consequence that the public, enamored with the immanence of this ever-present past and the heroes of other ages, conceives of science as something mostly dead and finished. It is lifeless, entirely logical, and generally dull. While we revere dead scientists, living ones are thought to be mostly weirdos. When young children are asked to draw a scientist, they invariably pick a dead one.

And it's not just the education system that could use reforming; it's the whole cultural zeitgeist. Popular books, magazine articles, TV shows—they all portray the science narrative as one of continuous incremental successes punctuated with a few brilliant leaps forward. All this may sound more enthralling, but I submit that it is actually a less engaging a story because there is no way for you to identify with it—unless you're one of those geniuses who always gets things right. The story is both more absorbing and more realistic if it is told with the stops and starts and the false ideas that seemed good at the time—that is, with all the failures along the way. Some of those failures were really great, tremendous, fabulous failures. Others were failures to perceive what now may seem obvious but just couldn't be seen at a particular time because humans . . . well, they just weren't thinking that way. Isn't it always fascinating to try to understand why, with basically the same information but the wrong perspective, some things might just as well be invisible. Which leads you inevitably to consider what is invisible to us today, hiding there in plain sight, because it is currently inconceivable.

Not permanently inconceivable, just currently so. What great failures are we in the midst of now? What cul-de-sac are we vigorously investigating? What simple truth is about to be revised? This is personal, living science.

Perhaps an example, a historical case history, would be valuable here. But immediately there is a problem. I can hope to be only partly successful in writing the narrative of failure, if just because many of the important failures have gone unrecorded. What we have are long blank periods that must surely have been filled with some activity, or periods totally dominated by a single authority in which other perspectives, ultimately right or wrong, were suppressed or at least ignored. There is the filter of the Middle Ages, an ecclesiastical period in the West that pointedly filtered which works were copied into the libraries. And works that were not actively preserved because they were not valued were likely to have been lost to mold and rot.

Remaining details are often sketchy. They are difficult to interpret, even by professional historians, which I am not. Four hundred years pass. What happened during all that time? Failures not worth recording? People got lazy? Complacent? Stumped? Are those things possible now? Could we moderns enter into a blank period? Are blank periods a necessary part of the process, an occasional feature—a long period of chasing after an ultimately wrong idea? Many wrong things can look right for an awfully long time. Slavery seemed fine for thousands of years. Vital forces and teleology were mainstream in science for millennia. And of course belief in magic and spirits of all sorts seems to have been with us since the earliest of times.

Well, there's nothing to do but work with what we have and make the best of it. Which is exactly how it's been done in science for hundreds of years—make an approximation; work up an imperfect model; look for someplace where progress can be made; accept, measure, and include the uncertainty; and be patient for an idea or finding that will emerge from all this tinkering and questioning. Here we go. Expect failure.

I'm going to use the circulation of the blood as my (non) narrative. It's an interesting example because how the blood circulates through the body is now such common knowledge that even little children know it, if only because they've been told. But its history is full of fits and starts and mistaken turns. It now seems so obvious that blood must flow around the body in some sort of circuit that it is almost impossible to imagine how it could have taken more than a millennium of scientific inquiry before William Harvey mostly figured it out. Perhaps even more incredible, it took yet another several decades after his ideas were publicly available before they were widely accepted. Harvey had to battle with wrong but entrenched ideas about vital forces and the pneumatic theory of the blood, as well as some plain old wrong anatomy "discovered" by the famed anatomist Galen, a Grecian scientist of considerable influence in the early Roman Empire. In a kind of parallel with Darwin, Harvey was sure about blood circulation as early as 1612 but delayed publication for more than 15 years, until 1628, for fear of publicly disputing the ideas of Galen. While it now seems obvious that blood circulates around the body, let's explore how improbable such a thought would have been to most anatomists for centuries.

First, there's the simple problem of complexity. You have an unimaginable 88,000 kilometers (55,000 miles) of veins in your body. That's enough to go around the equator twice! How amazing is that? Each one of us with all that blood vessel packed into our bodies. Now the big ones, the major arteries, are not so hard to keep track of, but then they branch and get smaller and smaller and finally they get so small you can't see them (the capillaries) without a microscope or some magnifying instrument. So there are lots of vessels and then they disappear into some sort of meshwork. Very confusing.

Confusing maybe, but nonetheless there were no shortage of theories, mostly wrong, about this life-supporting vasculature. The pneumatic theory of the blood, a particularly persistent one, paints blood as being a hydraulic substance that carried the imponderable life force. Silly, you think? Imagine if you will that you are a Roman soldier in the first century CE, on a battlefield in the cold early morning, the dawn lighting the sky just enough to obscure all but the brightest stars. Your fellow soldier is suddenly pierced in the chest by an arrow and falls from his horse. Blood pours out of his wound and onto the ground. In the cold morning, the hot blood begins to steam, and wisps of a weightless vapor from what appears to be his insides drift up and away, just as you watch your comrade's life drifting away. Wouldn't it be perfectly reasonable to see that vapor as a vital force, a spirit, a soul, that was fleeing the body and leaving it insensible and inert? How else to describe this ineffable vapor that correlates with the living state versus the lifeless one? You could have fooled me.

The pneuma theory was systematized by two early and influential physiologist-anatomists, the Greek Erasistratus, working around 250 BCE, and Galen, also born in Greece but who lived and worked in the Roman Empire around 150–200 CE. I cannot sufficiently emphasize the historical role these two men played—one could easily say that the sciences of physiology and anatomy began with them as much as Galileo is considered the first true physicist. They dissected human bodies, an activity that even today for most people is a barely thinkable thing to do. Human dissection, which has come in and out of favor and acceptability, is a deeply scientific kind of investigation. You can't help thinking "wondrous machine" when you open that hood. Galen, Erasistratus, and their associates (some of whom we know about and many of whom are unrecorded or unpreserved) published many detailed and thorough works on every part of the human body, as well as made comparisons to other animals. Between them they described most parts of the body, which still carry, to the continuing bane of medical students, the obscure Latin names they gave them. Among the hundreds, more likely thousands, of structures they identified were the aorta, pulmonary artery, cranial nerves, ventricles, auricles, and even the tiny but crucial cremaster muscle that controls the male scrotum.

For all of that, they were mostly wrong about how it all worked. This is a not uncommon feature of science. Technical dexterity, advanced measurements, reams of data—all precede understanding, sometimes by quite a while.

Galen and company were stuck on pneuma. Not hard to imagine that they saw the breath as all-important. They thought blood possessed and carried pneuma—or vital force—to all parts of the body. Pneuma entered through the breath, was added to the blood, which had been produced in the liver, and then was distributed throughout the corpus as vital spirit. If that all seems entirely made up, it's not that they had no evidence for this model. They did. But they failed to interpret that evidence correctly.

Erasistratus had noticed that the arteries in human corpses were empty, as they would be in a dead body, because the lungs stop moving before the heart stops beating and so all the arterial blood is pumped out with no incoming fresh supply. By the way, this is actually good evidence, used by

Harvey a millennium later, for a circulating blood supply. But to Erasistratus it meant that the empty arteries were the conduit of the airy pneuma. The vessels were thought to come from the liver, where the blood was produced. The liver is an especially bloody organ, so that's an understandable false assumption. Erasistratus observed the blood being drawn into the heart during diastole and then being pumped away. He ascribed to the right side of the heart the role of pumping the blood and to the larger left side pulsing out the pneuma. He observed the valves that prevented the blood—and the pneuma—from flowing back into the heart. These valves would become a key element of Harvey's circulating blood model. Erasistratus, in 250 BCE, was within a literal heart's beat of recognizing the circulation of the blood, but was prevented from that insight by his pneumatic theory. Remarkably it would be more than 1500 years before the idea surfaced again.

This was at least partly because scientists of this and the following eras let their philosophical worldview override their empirical observations. They were philosophers as well as scientists, seeing the perfection of the organ systems and the intricate workings of the body as an indication of Providence—they were in a sense early believers in divine design.

Galen, the Prince of Physicians, in his towering work *Uses of the Parts of the Body of Man*, sought to show that the organs were so perfectly constructed for their purposes that nothing better could be imagined, and this had to be proof of some sort of divine designer. Thus the aim of scientific observation was to determine the ultimate purpose of each organ and to understand the perfect way that it accomplished that function. He was, in fact, an early—perhaps the earliest—proponent of divine design, although it is unlikely he would ever have characterized it that way. Christianity and its providential beliefs had not really gained wide popularity yet, but Galen was preparing the way. Some historians believe this may be why so much of his writing was preserved by the ecclesiasticals of the Middle Ages—more so than almost any other pagan writer. And through that filter we have likely lost many other crucial ideas about physiology and anatomy. Of course it's always difficult to tally up what's been lost.

It was this "divine body" attitude that almost completely interrupted scientific investigation, for it invoked an authority against which there was no argument. Authority, and the infallibility that comes with it, is the bane of scientific discovery. When failure is not possible, neither is discovery. We have here an excellent example of a scientific narrative that is corrupted by its failure to record failures. I would like to tell you about all those failures as well but I can't. In its stead let's continue to follow the failure of what was the predominant idea for centuries.

Galen furthered the prominence of the theory of the pneuma, and his writings were so prolific and his authority so great that nothing much budged for several hundred years after his death. The simple failures of learned, hard-working, intelligent men like Erasistratus and Galen, because they were firmly attached to an authoritative worldview, were taken as indisputably correct. There was no possibility that they were riddled with wrong ideas, failures, which would have been far more interesting and informative. Without the possibility of failures to suggest new questions, a next generation of scientists never appeared. There were contemporary authorities, but they mostly copied the works of Galen into new textbooks. Galen's work brought progress to a halt for centuries. I think we often forget, in the foreshortened timelines of today's scientific pace, how long it was between discoveries in the centuries and decades up to as recently as 1900 or so. Erasistratus to Galen was 400 years. And that was fast compared to the next epoch.

Indeed, the next great steps forward in anatomy and physiology had to wait through the so-called Dark Ages until the Renaissance gave us Vesalius in the early 16th century. The historian Charles Singer makes the interesting claim that it was the confluence of Renaissance art, humanistic learning (wide availability of translated and printed classical writings), and a renewed enthusiasm for dissection that, perhaps ironically, brought life back into anatomy as a science. The timing is interesting because it was nearly contemporaneous with the appearance of what we would today identify as science-based physics. Historian Derek Price believed that it was the unique combination of Greek geometry and Babylonian numerosity, combined with a Western curiosity for astronomy, that created the perfect intellectual storm for the outpouring of physics beginning with Galileo. So in two important cases the development of science, be it physics or physiology, depended on the unique, and seemingly improbable, confluence of two particular cultural mindsets. Does it seem to you, as it does to me, that the whole thing is a bit of a crapshoot?

Vesalius (1514–1564) is the first modern anatomist, indeed perhaps the first modern scientist (as I've said, a place usually given to Galileo). Andreas Vesalius was the product of his day, created by the milieu of the Renaissance with its parallel reverence for the classics and revolutionary zeal for the new. Trained in Galenic philosophy and anatomy, the only tradition on the subject to have survived the long dullness of the Middle Ages, he was nonetheless restless with the classical methods of anatomy. When he took over the Chair in Anatomy at the medical school of the great University of Padua, he reformed the old system of a lecturer and his dissecting assistant and "put his own hand to the business of dissection." In short time he was a popular lecturer attracting large audiences to his dissections and demonstrations. By the time he was 28, in 1543, he produced the book that was to become the basis of anatomy for centuries, *On the Fabric of the Human Body*, sometimes known simply as the *Fabrica*.

To put things in some temporal perspective, Vesalius worked and published contemporaneously with Copernicus, and nearly a hundred years before Galileo. But what was a hundred years in those days? His *Fabrica* is not only the foundation of modern medicine, it ranks with Copernicus's *On the Revolution of Celestial Spheres* and Galileo's *Dialogue Concerning the Two Chief World Systems* as among the first books of true Western science. Knowledge was to be based on evidence and observation—not authority. The world, astronomical and biological, was to be understood on its own terms and not simply as the perfect expression of a divine vision with us at the pinnacle.

Vesalius's great contribution was his blending of art and close scientific observation—along with grave robbing and body snatching to obtain corpses. (At the time there was a law in neighboring Bologna more or less permitting human dissection as long as the body had been obtained more than 30 miles from the center of town. I guess it was rude to dissect your neighbors.) In one unfortunate case he is supposed to have begun performing an autopsy on a woman who was unexpectedly still alive! Unlike the anatomists of earlier days he drew his figures in lifelike poses, emphasizing how anatomy worked, not just how it was constructed (an idea he got from the gruesome living autopsy?). Vesalius repudiated the teachings of Galen, although his education was heavily influenced by Galen and he maintained some of the teleological approach to anatomy—that organs could be best understood by their purpose. For better or worse, a teleological approach actually works in biology—for a while. It's another of those cases where something wrong is nonetheless helpful. Vesalius taught that it was not the individual organs and tissues that mattered but the whole, integrated body—the fabric of the whole and thus his book title *The Fabrica*. Vesalius, it must be admitted, is a bit of scientific hero. He was one of that generation who dragged us out of the simple, settled, complacent medieval worldview with that ability, still amazing to me, to look at the same things as everyone else and see something different.

This change in attitude and the use of empirical data against authority was the basis of an entire school of anatomy that followed Vesalius's principles. Among its greatest members were Fallopius, who gave his name to the eponymous tubes down which eggs travel to the uterus, and one Renaldo Columbus, no relation to the explorer. Columbus made the important discovery that systole, the cardiac cycle of contraction, was synchronous with arterial expansion, and, conversely, diastole or cardiac expansion was timed with arterial contraction. For centuries the opposite had been the belief: that it was the expansion of the heart (during its filling with blood) that forced it against the breastplate and gave rise to the sound of the heartbeat and was therefore the important part of the cycle. It is of course exactly the opposite—the muscular contraction of the heart is what we now understand to be the heartbeat. An easy enough mistake to make, but one that completely reversed the mechanism of the heart, making it a suction device instead of a pump. Without Columbus's simple correction it was impossible to understand the blood as a circulating fluid, being pumped out into the body through arteries and passively returning through veins. Somehow this Columbus's discovery did not fare as well as his mariner namesake's and it was forgotten—or ignored—for another hundred years until William Harvey rediscovered it.

Harvey changed everything. The time is the period from 1600 to 1630—contemporaneous with Galileo, but some decades *before* Newton (who was born in 1642). Harvey also spent time at Padua, the great center for anatomy established by Vesalius. Indeed, he was a visitor there when Galileo was giving his famously popular public lectures. What he brought back to England, in the early 1600s, was a new passion for comparative anatomy—the understanding that other animals were mostly similar to humans, even if different in certain details. After cogitating for some 20 years he finally published his great work, *An Anatomical Dissertation on the Movement of the*

Heart and Blood in Animals, in 1628. This 72-page publication, barely more than a pamphlet, transformed the scientific outlook on the human body. Harvey had integrated physiology with anatomy. No longer was the body merely a parts list. Structure was melded with function and integrated with the workings of the other parts of the organism. It was Galen and Vesalius, but without the teleology or theology. As Copernicus and Galileo removed man from the center of the universe, Harvey showed us to be more of a machine than a divine creation. Science was providing a radical new worldview, and the genie was done with the bottle.

The actual steps and the experiments that Harvey made to reason out the circulatory nature of the blood are a wonderful science lesson, although too long and detailed to go through here. The key insights, however, were so simple as to make one wonder how it could have taken over 1500 years to come to them. The first is that the blood passes between the two sides of the heart, one side (the right) connected to the lungs (or pulmonary system) and the other to the whole rest of the body. This idea, and evidence for it, had been around for a hundred years, but was ignored, and even suppressed, because it was inconsistent with Galen's views and offered no place to put the all-important pneuma (remember Erasistratus saw this, too, almost 2000 years earlier). Once Harvey rediscovered and accepted this crucial idea, the question was, where was all that blood coming from? Galenic dogma claimed that blood was produced in the liver and lost through invisible pores to the organs and through the skin. In a simple calculation Harvey estimated that the ventricle holds about 2 ounces of blood. If the heart beats 72 times per minute, then in one hour the left ventricle will pump no less than 8640 ounces into the aorta—a remarkable 540 pounds of blood, more than three times the weight of an average man! Even Lady Macbeth was stunned by the amount of blood that flowed from Duncan's fatal wounds: "Yet who would have thought the old man to have so much blood in him?" (*Macbeth*, Act V, scene 1). There was, of course, only one explanation—it was the same blood going round and round.

It is hard to convey the importance of this insight. It is so simple you could imagine about a hundred earlier scientists slapping their heads and wondering, "Why didn't I think of that?" But then it is also so revolutionary that it changes completely and forever the way we think about the human body. In this it is no less remarkable than Copernicus seeing that the Earth went round the sun, or that Galileo and then Newton could work out the concept of inertia. Overnight Harvey destroyed all vestiges of the millennial ideas of vital forces and pneumas. He ushered in the age of rational inquiry and experimental physiology as a hallmark of studying living systems. For the first time life was not only the province of theology and philosophy; now it belonged to science as well.

Except it didn't happen that way. It should have been that Harvey's work swept away all the claptrap that preceded it, but it took another couple of decades for the circulation of the blood and its consequences to be widely accepted. "I'd rather be wrong with Galen than right with Harvey," a group of influential physicians of the time reportedly said. Some form of vitalism continued to dog biological thought through the 1800s, and bloodletting was a favorite medical treatment for almost every conceivable ailment until at least the late 18th century. Measuring the pulse and blood pressure did not become common medical practice for another *two centuries!* Even ideas whose credibility should be immediately obvious can evade the fog of human intelligence.

For that matter, where are we with blood today? We can't decide if cholesterol is good for us or not, or what the right level of it is. We do angioplasties to clear out arteries, but there is debate about their lasting value versus the more radical bypass surgeries. Blood pressure is now measured obsessively, but we still don't know the exact range that is acceptable or what the differences in blood pressure between sexes, races, and ages mean. How long did it take us to understand the dynamics of a blood-borne disease like AIDS? Only recently have we recognized that the blood is also a major immunological organ. Indeed, only recently have we come to look at the blood as an organ. How nice to see how much we still don't know.

I apologize for this perhaps overly long narrative, but I had two purposes in mind. One was to provide an alternative to the very overworked story of physics from Newton to Einstein as the ultimate narrative of Western science. There were other things going on as well. There was a scientific revolution, and it was not just in physics. The second, and more important purpose, was to show that the narrative of any discovery is not so straightforward as the textbooks and encyclopedias would have you believe. This one took nearly 2000 years and, contrary to my purpose but by necessity, I mostly covered only the major figures. There were dozens of others, some contributing correct data but wrong ideas, some adding wrong data, some insisting on one way over another for religious or philosophical reasons, and others trying to look just at the data but through crooked lenses shaped by the centuries of thought preceding them. There were long blank periods when virtually nothing happened. Many cul-de-sacs, detours, dead ends, and, even more frustrating, many correct ideas that were for all sorts of reasons ignored and forgotten. It is a process of piling good failure upon great failure, adding up to an insight that is easy to underestimate from our current vantage point, but that eluded the best minds for centuries. And this story is not special insofar as the narrative is concerned. It is what is going on right now in laboratories all over the world. In the sciences of climate, cell biology, physics, chemistry, mathematics—everywhere—mistakes are being made at an amazing rate. And progress is the outcome.

EIGHT

The Scientific Method of Failure

Success is advancing from failure to failure without losing enthusiasm.
—*Not* said by either Winston Churchill or Abraham Lincoln

In *Ignorance* I had not a few, and mostly unkind, words to say about the Scientific Method—particularly about the idea of the hypothesis. I decried the Scientific Method as a comical concept that no real scientist ever really practices and is taught only to schoolkids, presumably to make science look as uncreative as possible. *Method* implies that there is a rule for doing science, a recipe to be followed, as if it were a machine that just cranked out discoveries. This is both a pervasive and thorough misconception of science. I have become a bit notorious for claiming in a TED talk that what scientists really do instead of following this orderly, methodic procedure is more like "farting around." What I mean by this of course is not idle and pointless goofing off. But I do mean dabbling, fiddling around, puttering, tinkering—something akin to play. It's serious, but it's not very constrained.

Even more, I have ranted against the idea of the hypothesis, step one of the Scientific Method. This idea has been around the longest and causes the most trouble. My concerns with the hypothesis are that it leads to bias, in fact is generally not where science begins, and has been abused so terribly by government granting agencies and educational curricula that it would be better to just dump the idea all together. Most scientists have stopped using the word *hypothesis* in favor the more modern sounding *model*—as in, "Our *model* for human effects on climate predicts a raise of the planet's surface temperature by x degrees in z years." A scientific model is more or less synonymous with a *theory* or *hypothesis*, but somehow seems less a finished thing, more a work in progress, fluid, provisional, and in need of refining.

Even if we accept the Scientific Method as some sort of general description of how science is meant to proceed, it is of little practical help. The steps are as follows: (1) observe; (2) form a hypothesis; (3) design an experiment that manipulates the hypothesized cause and observe the new result; (4) update the hypothesis based on the results, and design new experiments. And then, as the shampoo paradox states (lather, rinse, repeat . . .), do it again . . . and again. This all sounds good, except that no scientist that I know of actually follows this prescription. It was originally developed in its more or less modern form by the famed empiricist Francis Bacon, who in an ironic twist died as a result of following the method, perhaps a warning to future scientists. During a journey through a snowstorm he was struck with the idea of showing that the Scientific Method could be successfully applied to the problem of the preservation of meat. He hypothesized that cold temperatures would preserve meat longer. Stopping to collect ice and snow to stuff a chicken carcass, he came down with a fever and died less than a week later from pneumonia. According to a letter he wrote, just before succumbing, the experiment was a success. So there was that consolation.

All well and good, except nothing is said about how to actually come up with a hypothesis. Verification of the so-called hypothesis is the most pedestrian step in the process, instructions for that being needless. On the other hand, the most critical step in the whole cycle, the one that requires a magic brew of creativity, thought, inspiration, intuition, rationality, past knowledge, and new thinking—this the Scientific "Method" has nothing to say about. "Form a hypothesis." Very good. *How*, precisely, does one do that? Do hypotheses just come in a catalog and you choose the one that looks best? What if there seem to be two or more hypotheses—how do you choose among them? How do you decide that one hypothesis is more reasonable than another? This is like giving an art student a brush and the direction to "do painting." Bacon believed that meat could be preserved by low temperature. We don't know why he believed that; there was no body of knowledge at the time that would lead one to that conclusion. Microbial organisms were unknown and unimaginable to Bacon. He could just as easily have thought that heating meat would preserve it—as indeed smoking does. So the fatal experiment was the result of an idea that belonged entirely to Bacon's mind—it was not yet supported by any but the most casual of observations.

Ironically, this is much closer to how science actually works. Not as Bacon imagined it, but as he did it. This is best characterized, as I have said before, by *not* following the rules of any investigative method. It is a state of mind that is best suited to *making* problems—not just solving them. And then identifying the right problem among many by using intuition, instinct, perception, talent, irrational impulse, and of course knowledge, to identify some mystery, some uncertainty that gnaws at you and motivates you to try one crazy thing after another until something gives you a glimpse of a possible solution. It is the creative part of science where there is no recipe or method that instructs you how to simply deposit data, turn the crank, and get results. It is perhaps best described by an ancient proverb that I quoted to open the book *Ignorance*: "It is very hard to find a black cat in a dark room. Especially when there is no cat." Stumbling around in dark rooms in search of black cats that may or may not be there is the best description of everyday science that I know.

So in the end the Scientific Method is more dangerous than just being a quaint approximation of what scientists do—it has that unfortunate trait of seeming to say something when it really says nothing. The result of those sorts of formulations is that everyone is satisfied with the state of things—it's been explained, it gets into the textbooks, it's what students learn and can be tested on—but it's not true or correct or even approximately so. It's a calamity, this "Method."

What should it be replaced with?

The first option to consider is "nothing." It doesn't need to be replaced because it, the Scientific Method, wasn't really there in the first place. It was an abstraction, a shorthand, an impostor, a description of something that no one used. For most scientists, I think that solution would be fine. Why replace something that was never really there to begin with? Is it even possible to have a simple, unique, formulaic description of how to make science work? Perhaps the best policy is to mostly leave it alone. Not try to be too attentive or too nurturing, or too pedantic. Make sure it gets fed and kept warm and then let it out to play—as often and for as long as you can.

But I fear that won't satisfy the public or the funding agencies or the popular science press or the curriculum czars. And if we don't replace this pervasive, if empty, concept with something, they will just go on using it because it's been there for so long and is an easy shorthand, and it's less cumbersome than some longwinded description of the actual process—which no one may agree on anyway.

So at the peril of bringing down the wrath of both my scientific colleagues and my trusted friends in the philosophy and history of science, and who knows who else, I will make some hubristic attempt to carve out an alternative to the Scientific Method. Something a little more precise than just screwing around—but not much more.

Let's start from "stumbling around in a dark room" and see if we can build on it. So you begin by messing about, trying stuff, even stuff you think has only a small chance of working. Not just any stuff, of course. Often it originates with something you may have observed or read about or heard in a seminar—something that you can't explain. Scientists are driven by a desire, perhaps a need or obsession, to explain. Explanation is the best kind of understanding there is. There are many other ways you might "understand" something—religiously or spiritually, morally, ethically, legally, socially, even intuitively. But understanding by explanation is particular to science.

Science understands something by having an explanation, and a particular type of explanation that is different from the explanations provided by those other approaches. A scientific explanation not only tells you why something happens the way it does, but how, and why, it is likely or not to happen again in the future, given certain conditions. Note that the scientific explanation does *not* attempt to provide meaning. God, they say, is unpredictable, or at least we cannot know his mysterious ways; morality and laws differ from culture to culture; intuition, well, intuition maybe it works under certain circumstances, and maybe it doesn't. These are all explanations, but they are no better (and much less entertaining) than Kipling's *Just So Stories*. They don't explain in a way that can be applied without regard to human social foibles and biases. They are not independent of everything except observation.

The cosmologist and very smart thinker David Deutsch, in a book that is, for better or worse, bigger and deeper than allows for a casual reading, has put forth this idea of explanation better than anyone I have seen. I am borrowing heavily from him in this discussion. I recommend his book *The Beginning of Infinity*, but I advise you—it is a commitment.

Let's take explanation a bit further if we can. One thing we can observe about *scientific* explanations is that they are always being revised. I would make that part of the definition of a scientific explanation: it can and will be revised. This seems at first to undermine the idea of explanation—after all, an explanation is an account of how and why something happens or happened. If it's going to be revised—that is, if it isn't the whole story—how much of an explanation is it, really? This is precisely the magic of scientific explanations. They can be, and usually are, correct even when they are incomplete. Deutsch claims this is because a good explanation, a good scientific explanation, is very hard to change. Whatever it did explain well must be accounted for in any new or revised explanation. So the explanation can change as new data requires, but you can't, and shouldn't, just throw the old explanation out.

I think this is at the base of the common misconception that many people have about the famous paradigm change model introduced by the historian Thomas Kuhn. This is the one where science progresses by bumping along incrementally until it begins running up against inexplicable bits of data here and there, and when that inexplicable data reaches a critical mass, then there must be some major upheaval, some tectonic shift of the fundamental paradigms, to a new set of explanations. The classic example of this is the shift in physics from Newton's mechanics to Einstein's relativity, from absolute time and space to relative frames of time and space. But even Kuhn admits that the idea has become an overstatement. Einstein has not proven Newton to be wrong, only that there was an explanation that could be more expansive and would be a closer approximation to the way the universe seemed to be. It's not as though 250 years of physics and engineering based on Newtonian principles were all wrong and we had to start all over again. Newton didn't land on the trash heap because of Einstein—ask any high school physics student. But there was certainly a paradigmatic change in the explanation.

Many times these various paradigms or systems of explanation can coexist, even though each one has some shortcoming as a total explanation. Hasok Chang of Cambridge University makes the case that the GPS device we are all so familiar with comfortably embodies four different scientific systems. It uses atomic clocks that are the result of the quantum physics model of atoms, in satellites placed in orbits calculated by Newtonian mechanics, requiring a low gravity relativistic correction with Earth-based clocks—all in the service of providing us with a map of a flat Earth!

Of course there are bad explanations, but these generally disappear, at least in the scientific literature. Sometimes they hang around longer than you might expect, because they are pretty good explanations for a while. And then they fail, without any hope of revision. Phrenology, the science (it was considered a science at one time) of determining personality traits from the bumps on your head, is one such explanation that didn't stand the test of time but did have an important impact in neuroscience. It was the first time that the brain was proposed to be the main if not sole source of thought, emotion, and personality (instead of the heart, gut, pancreas, liver, etc.). Further, phrenology considered that many of these traits were localized to particular parts of the brain. That is, certain behavioral or cognitive traits arose in specific locations in the brain. A stronger or weaker instance of one of these traits was reflected in an increase or decrease in the size of that part of the brain. This in turn would appear as a larger or smaller bump on the overlying skull. Since everyone had slightly different shaped skulls with different bumps, you could examine a person's skull and make predictions about their personality. Of course it was all nonsense. It was a bad explanation on almost every account. But it contained two new and correct principles of neuroscience—the brain is the source of personality, and its functions are largely localized. These two principles are now considered fundamental to brain studies and have given rise, however indirectly, to modern-day EEG and fMRI studies. And anthropologists are able to make interesting speculations about brain function in early hominids based on casts of their skulls. So even bad explanations, a pretty complete failure in this case, can have some value.

All this is to say that *searching for an explanation* may be a better description of what science does than any Scientific Method. How are they different? The Scientific Method suggests a very dispassionate, detached, third-person sort of search. As I noted, it doesn't really tell you how to form a hypothesis, just that you sort of passively make one up, sometimes out of thin air. You have objectively (as if there were such a thing for a human) observed something, and now a hypothesis forms. None of it really has much to do with you; after all, this is empirical science. The less it has to do with any individual the better. Of course we would like science to be impartial and unbiased, untainted with personal involvement. But this is an ideal that is not finally attainable, and maybe should not be. Value-free science may be an impossible goal, but we can fully recognize our partiality and hitch that to a set of procedures aimed at removing as much bias as possible. On the other hand, if you insist on proceeding dispassionately you will have left out a critical ingredient—passion. Dispassionate is too easily mistaken for objective.

Searching for an explanation begins more appropriately with curiosity, and curiosity is not dispassionate or impersonal. I think I can say with general agreement of my colleagues that it is deeply personal to the point of being idiosyncratic. You are curious about why flowers have colors, and while I don't care much about that, I am obsessed with how they have different odors, both good and bad. You are fascinated by tiny little lights in the night sky that are so far away you could never hope to approach one, while I need to explain why bacteria that can't be seen without powerful microscopes fill our guts. None of these pursuits is dispassionate—or perhaps even rational, if you think about it. They are all crazy things to worry about. But every one of those interests has led to cures and technologies that have improved our lives, and, more importantly, they have satisfied a common desire to know about the place we inhabit. Is there a better image of science than the guileless, but personal, curiosity of a field biologist collecting odd bugs and picking his or her way through some remote ecological niche? Or of a devoted researcher like Jane Goodall so deeply involved with her chimpanzee subjects that she almost became a member of the troop?

In science the pursuit often commences with failure—an apparently failed experiment puzzles you, or you notice a failure in some existing data, some inconsistency arises between sets of results reported from different laboratories, or new data calls into question some older established data. In other words, there is a failure, even just a small one, in the explanation. As the Leonard Cohen lyrics say, "There is a crack in everything / That's how the light gets in." A crack, a structural failure, as the engineers would call it, that's where a sliver of light shines through. That's where you direct your curiosity. Where failure reveals your ignorance and gives rise to curiosity—that's where the science gets started.

There are endless workshops on how to be creative, how to release your inner creativity, how to bring creative solutions to the job and the marketplace, and all the rest of that bull. Here's the bottom line—no one knows the neurobiological or even the psychological basis of creativity, imagination, or curiosity, much less how to induce it reliably. The typical workshop tactic is to try to figure out what creative people do, and then reverse-engineer that into your behavior. One thing that comes up regularly is that creative people are creative by putting things together that don't normally appear to belong together. They can work across categories to find novel solutions. That's an interesting post hoc observation, but hardly a prescription for actually being creative. After all, first you still have to have ideas, and now you need them in many different and disparate areas. And no one can tell you how to do that.

In science, and I think in art, creativity comes from failure. Not from putting things together, but from seeing them fall apart, even making them fall apart. Creativity arises in discrepancy, in the breakup of things that have been thoughtlessly joined for too long. In that space, of *not* knowing and *not* understanding, creativity can occur. That space is the void of failure and ignorance. Where else can new ideas come from? They are not coming from things you already know—except insofar as those things have created new questions. When you try the obvious solutions and they don't work, the failure forces you to open your mind to alternatives.

It is just as likely that creativity is best measured by the capacity to dissociate ideas that are traditionally inseparable. It costs more to dissociate ideas than to associate them, but at least you have a place to start. How better to dissociate the inseparable than by failure? New ideas come from the unknown and the unknown is where failure rates are highest.

This is the way of science. Continuous searching and unending curiosity rightfully replace simple procedural formulas. Science is never finished, and that serves to increase its value.

The word *debacle*, which we use now to mean "unmitigated disaster or total failure," originated from the French *débâcler*, which means literally "unleashing." It was originally used in French to refer to nautical ice breaking—that is, breaking up something solid so as to provide new pathways. It entered into English usage in the early 19th century, but how it got its currently negative connotation is not known. I think we should return to the curious double meaning buried in its etymology—unleashing and failure—breaking things up to reveal a new path. I guess we could use the word *creativity* for this as well.

. . .

The famous and very thoughtful geneticist François Jacob made the distinction between *night science* and *day science* in his wonderful book *Of Flies, Mice and Men*. These are such wonderful phrases because it is immediately clear what these two aspects of the same activity are: day science is the logical, rational—yes, even methodical—pursuit of data; night science is the intuitive, inspirational, any-idea-is-worth-entertaining pursuit of discoveries. It is in some ways, the distinction between the romantic and the empirical, but on a personal level. Day science may have some little bit to do with the Scientific Method as it is commonly stated, but night science is another creature altogether and it follows different rules—or no rules at all.

Of course both are part of the enterprise, but we all know that the adventure, the real leaps, the advances that matter most—they are the stuff of night science, the "Aha!" moments that appear most unpredictably. The solitude of the late night, the absence of a clock—once the sun sets, the whole night looks the same until first light, unlike the day, which splits easily into morning, midday, and afternoon and has activities specific to each. Nighttime thinking seems closer to dreaming than daytime thinking. There is less urgency, less direction, less focus as things float in and out of mind. It is conducive to curiosity.

But these are just metaphors. Night science is not necessarily restricted only to the time after the sun sets. It can be done anytime that the mind is in the right condition. Solitude and darkness are not requirements; they are just ways of helping it along. Night science can be done in the brightest of lights, and in the chaos of a big lab humming along during the middle of the day. It can, and is, done after each experiment that delivers a puzzling result—the "Hmmm, that's strange," which Isaac Asimov famously identified as every scientist's favorite phrase when looking at data.

And what method governs this kind of night science? Why, none, of course. So why do we teach and worship the

Scientific Method that is at best only a pale description of what we might do during daytime science—and barely covers that? I'm sorry to disappoint, but my hubristic attempt to replace the Scientific Method has failed—and I'm thankful for that.

NINE

Failure in the Clinic

Medicine is a science of uncertainty and an art of probability.
—Sir William Osler, founder of Johns Hopkins Medical School

Failure is especially difficult when it comes to medicine, the consequences being potentially disastrous. The physician's practice of medicine is not, strictly speaking, science, or not science alone. But the technology, the procedures, and foundations of modern medicine are very much rooted in science. Like technology, then, it is often useful to think of it in scientific terms. So what can we say about failure in medicine?

Our very earliest ancestors were likely to have been preoccupied by medical concerns. These concerns may even have predated interest in the heavens, making medicine or some form of it, rather than the oft-cited astronomy, the most ancient protoscience. Early humans must have been active in discovering, possibly by accident but surely at least partly by trial and error, herbal treatments for toothaches or infections, methods for managing birth and death, and surely intoxicating substances (modern-day gorillas are known to select fermented fruits for their inebriating effects and reindeer mushrooms for their hallucinogenic properties). Though likely quite crude, medicinal treatments were possibly among the first kinds of information passed down between generations, and maybe the first to be practiced by "professionals."

Given that the placebo effect works even in animals, these crude medical practices were likely to have had a dependable success rate of something better than 30%, the generally accepted level of placebo effectiveness. That is, around a third of the population receiving the placebo drug or treatment will get better. This is why FDA testing guidelines require that, depending on the particular controls being used, a new drug or treatment result in an improved outcome in more than one-third of the patients receiving it. There are documented cases where placebos have been more than 60% effective. Even among primitive humans practicing the most rudimentary medical interventions, the failure rate might have been deceptively low. Most shamans would make a decent living curing 33% of the tribe's ills, and no doubt believed in the procedures they were using.

Placebos are one of the most confounding aspects of failure in medicine. From the discovery of new drugs to the valence of bedside manner, it confounds failure with apparent success. Because this is medicine, there is an ethical component as well. After all, if the placebo works, if the patient improves, shouldn't it be as much a part of the medical toolbox as the expensive drug or procedure? If the attentiveness of the physician is enough to cure the ailment, then shouldn't that be the first thing to try? (I have often joked that I prefer to have the placebo—you get all the benefits and none of the side effects. Although that's not really true since placebos can, under the right conditions, produce the expected side effects also.) On the other hand, is it proper to pursue a treatment you know to be useless from a purely empirical scientific perspective—the pill that has only inert ingredients or the fancy machine with panel lights blinking to look like they're doing something?

How the placebo effect works remains somewhat mysterious. While numerous studies, both large and small, have established the reality of it, none have actually determined the cause. In vague terms it is clearly psychological. Like hypnosis, some people are more susceptible to it. But don't confuse susceptible with gullible. Often people who claim to be too sophisticated to be susceptible to placebo effects (or hypnosis) are nonetheless among those for whom it works. Indeed, the placebo effect extends to the doctors administering the treatment: if they believe they are giving a real drug, the outcomes are better than if they don't know whether it is the placebo or the experimental treatment. Doctors are themselves sometimes placebos. This is why most studies are conducted under "double-blind" conditions—neither doctor nor patient knows which treatment is being used.

From a scientific perspective, the placebo effect is a nuisance. It is of interest to study on its own, especially as it reflects on the poorly understood interactions between the mind and the body, and there is a considerable current research effort directed at understanding it. But because it interferes with failure, making it look like success, it complicates learning why a particular drug or treatment does or does not work. For the selfish purposes of this book, it is a perfect example of why a good, dependable failure is so important.

Let's change gears a bit here and examine true failures in medicine. As important as failure is to the practice of science, it is usually a terrible outcome in medicine, and hardly desirable. Nonetheless, it happens, and understanding it and using it are no less important in the daily practice of medicine and for the development of its tools and treatments than in any other area of science. One roadblock here is the legal, and therefore financial, ramifications that may accompany medical failures. Here, as elsewhere, we must be careful to make the distinction between careless error and true failure.

Even the way failure is measured in medicine is different, although it confusingly uses the same phrases as normal probability. When told a procedure is 95% successful, the patient may not realize that this means that for 1 of 20 people

it will be a 100% failure. This is quite different than something that, when applied, is 95% successful. These are very
different things that alter where we would look to improve
outcomes or how a patient and doctor should come to a
decision about using a particular procedure or drug. It is a
question of accurately understanding failure, and accurately
communicating it.

I interviewed a wonderful doctor whose confidence I will
maintain here. Suffice it to say that she—yes, she—is a leader
in the field of orbital surgery. No, she does not work for
NASA, although even she would admit that kind of orbital
surgery specialty would be very cool. Orbital surgery is very
delicate, very critical, very focused surgery that repairs what
is commonly known as the eye socket and the surrounding
bones. It is a mix of cure and cosmetics—repairing what
keeps your eyes in place is critical to your vision, and usually
the problem would be severely disfiguring without the care
of a plastic surgeon. Orbital pathology is so specialized that
to her the eye as just that thing that fills up the space she
is working on. Her main operations involve tumor removal,
orbital decompression for thyroid disease (which causes bulg-
ing eyes), and reconstruction following orbital trauma—what
she classifies in males as typically caused by "the fist of the
new boyfriend of the ex-girlfriend." Dr. I, as I will call her,
has no time for incompetence in medicine and especially in
surgery where real damage can be done, so often in the name

of healing. This, she believes, is in some small part due to the unethical money-driven incentives in medicine and also in some part due to the ego that seems to be a prerequisite for those who become surgical specialists. Failures from these causes are unforgivable. But there are deeper sources of failure that are perhaps endemic to medicine. These failures must be forgivable if we are to make progress. It is on this razor edge of making medical failures useful that Dr. I has a nearly unique experience.

Dr. I belongs to a remarkable, apparently one-of-a-kind, club. It has no name or website, has only 40 members, orbital surgeons from around the world, all of whom are invited for life, and holds a closed meeting once per year at which attendance is absolutely mandatory. Miss a meeting, and you're out of the club. And at this meeting they discuss failures. Each member is expected to come prepared with a presentation of failures they have had the past year. From a single case to a few cases. Failures of diagnosis, of technique, of treatment. Personal failures and professional failures. Unlike a normal conference, you don't make your presentation and then take a few questions from the audience. You are constantly interrupted, queried, asked for details, alternatives, what you were thinking then, what you think now. Mistakes are expected, and confronted. The audience, the club, is composed of the leaders in the field. As Dr. I notes wryly, your audience is your library—every textbook is sitting there in the audience.

Finally, the aim of the club is not merely to self-castigate, but to spread the knowledge collected in this small and special gathering. The club is designed to probe faultless failure, something that you could not do in a larger public forum. But its true value becomes evident when members leave the meeting and write papers, revise textbooks, lecture, or consult.

Every major hospital now holds a weekly meeting known as the M&M, M&M being Morbidity and Mortality—that is, injury and death. These often tense assemblies are not entirely unlike those of the club that I have just described, although Dr. I's organization predates regular M&M meetings by a decade or two, even though the idea has been around, and often resisted, since the 1900s. There are important differences as well.

In the hospital M&M meeting, the medical hierarchy is on display. The most senior surgeons sit in the front row; chief residents present cases that have either resulted in a patient's death or included some other potentially life-threatening mistake. The descriptions are in the passive tense—no one screws up: rather, a procedure was attempted without success; "anesthesia," not a particular person, "was able to place an airway tube." The responsibility rests entirely with the attending physician for the case, whether or not he or she was even in the hospital at the time of the incident. It is the attending's job to supervise every case; he or she is the doctor on the line. This person may delegate a job to a resident or a nurse, but he or she is ultimately responsible and answers the final question: What would you have done differently? Typically the answer has to do with some procedural issue involving having the right personnel in place at the right time, and a note is made to improve on that, and then the proceedings move on to the next case.

I don't want to make it sound as if these meetings are just a show. They have had a significant effect on medical practice and on ferreting out areas where mistakes were being made needlessly. But they do not, in general, challenge accepted practice—just the manner in which it is being carried out. Dr. I's club is different in this important regard. First of all, it is personal and not hierarchical, and it is active and not passive. Ironically, it is easier to be open in a closed group than in a so-called open meeting. "I did this and it didn't work." She says this with sympathy and compassion, but also factually. Questions are asked of the actual individual who did the procedure, and it is their insight into why it didn't work or why it went wrong that is thus made available. Techniques, not just competence, are questioned. Methods of practice that have been accepted for years are called into question for specific cases. Could this have been done differently? Should this have been done at all? Dr. I claims that a frightening number of cases are of the sort where doing nothing would have been the better choice. So why wasn't that chosen? Could it have been known beforehand that doing nothing would have been better? In the future, how can you recognize when doing nothing would be better? How can you convince the patient that doing nothing is the best option? The old adage in medical practice *primum non nocere* (first, do not do harm) is often neglected in this interventional era.

Even incompetence can be occasionally positive—it brings you humility, resets your ego, resuscitates your vigilance, and finally increases your confidence. This may seem paradoxical, but Dr. I describes seeing surgeons going into a procedure with trepidation because of the perceived pressure of having to meet expectations that they recognize as unrealistic. Managing medical expectations, not only for the patient but also for the physician, is critical, and at least one way of accomplishing this goal is maintaining a proper attitude toward failure.

I don't want to dwell too long on medicine because it is so fraught with moral and ethical issues. Not to mention the economics that has come to drive so much of it. I refer not only to physicians, but even more to the entire medical infrastructure—from hospitals to insurance companies to device and instrument makers to the drug business. Medicine is a science, an art, and I'm afraid now an industry. It clearly deserves its own book on failure, but I am not the person to write it. Having had the opportunity to interact with Dr. I, however, it seemed remiss not to at least tell that remarkable story.

TEN

Negative Results

How to Love Your Data When It's Wrong

> If a machine is expected to be infallible, it cannot also be intelligent.
> —Alan Turing

So you have an idea, and you do an experiment to see if it comes out the way you think it should if your idea is correct. This is more or less the Scientific Method that you are taught about in school, and which no one actually follows. But let's say in this case you do more or less follow the method and do the experiment you hope will show that something works the way you hypothesized it would. And then it doesn't. Bummer. You do the experiment again, which is a really boring thing to have to do, but clearly you made some mistake somewhere and you just have to concentrate a little more this time and make sure you get all the steps right. So you redo the experiment, and damn if it still doesn't work. You check your notebook to see that you did everything the way you should have and it all seems in order. Definitely time to go have a coffee, assuming it's not late enough to go have a beer. This is clearly just some technical problem, either with you, although you've been careful, or with one of the reagents—the fancy name for ingredients. Maybe the salt concentrations were wrong? Who mixed that buffer anyway? The new undergrad work-study student in the lab? He could surely have screwed something up. Gaining new resolve, you decide to mix up everything fresh, from scratch, and do it yourself. Back to the bench.

Maybe it works finally. Or maybe it never works because you had the wrong idea. If it works it may become a data point in a figure in a paper, part of a statistical analysis, or maybe even a figure of its own if the result is central enough to the story. But if it never works then typically it gets buried in a lab notebook that no one will ever read through, even you. It may come up in a discussion at a conference, at the bar over a beer when someone says, "Hey, I wonder if so and so . . . ," and you say, "Nah, I tried that and it never worked." This is what's known as a negative result.

Most of the time experiments don't work for one reason or another. So most of science is negative results. But unless it happens to come up at the bar at a certain meeting on a particular night, who would ever know about it? You, your closest friend in the lab, maybe a few others if it got discussed at a lab meeting, your advisor. That's the furthest it will ever go. It would appear that the vast majority of the science that goes on in the world's laboratories never sees the light of day. I'm guessing you think that seems somehow wrong.

The issue is even more striking when it comes to research involving testing drugs or treatments in what are typically known as clinical trials. To be honest, I don't really think of clinical trials as science, but that's my own bias. In some ways, they are probably a purer version of science than basic research is. Clinical trials are highly organized, have lots of controls, are hypothesis-based, use sophisticated statistics, cost lots of money, are done by people in white lab coats—they have all the trappings of science. Basic science is just about poking around most of the time. It's what I call discovery, whereas clinical trials are about measuring accurately to find something out. I know they sound similar, but think about it a bit and you'll realize that they are barely related. One seeks to understand, and the other seeks to see if something works, even if the way it works is not entirely understood. We'll come back to this later; let's stick with negative results here.

When clinical trials fail, or produce negative results, then the stakes are a bit higher. Those data do have to be published (although they were not routinely published until recent legislation began requiring it—see Ben Goldacre's recent exposé of drug company subterfuge). They must be published even though they don't show what you wanted them to show—or worse, they don't show anything at all, which makes people wonder why you tried this to begin with. That's not entirely fair; many things look quite promising right up until the last step, and then they fail. We have cured cancer in mice over and over—no mouse need ever die of cancer. But virtually none of those cures have worked in humans. No one knows why. So we keep developing promising treatments for cancer in mice, and they keep failing in human clinical trials. But what's the alternative? One of them will work one of these days, and then perhaps we'll have some idea of why all the others failed. More importantly, but virtually uninvestigated, is why this is the case (this is one of the differences between discovery and measuring). I bet we could learn some interesting things about cancer if we went back and looked carefully at why some treatment worked in a mouse and failed in a human. But to my knowledge no one ever does that. A failed clinical trial is such a disappointment—the loss of money and time and hope is so emotionally draining that people just want to be done with it. And who can blame them? With all that money and time down the drain, do we really want to invest more to find out why the damn mouse is still thriving?

Lately there have been some questions about whether all the negative data from clinical trials get published, or if some are strategically withheld to increase the likelihood of FDA approval. Since the sums of money are often enormous, there is plenty of motivation to fudge things here and there. The recent case of Vioxx has become, in record time, a virtual historical case study. The serious side effects apparently showed up in the early weeks of the drug's use, and maybe even in the trials, but the data were suppressed, or at best ignored or discounted. Why this would be seen as a reasonable course of action is hard to know, since the inevitable outcome was that as the drug was prescribed for increasing numbers of patients, the negative results would come out—only they would be in the form of debilitating strokes, heart attacks, and deaths, followed by lawsuits. You wonder what people could have been thinking. It's true that every drug has some side effects. The First Law of Pharmacology is that every drug has two effects—the one you know, and the other one. Nothing is risk-free, but if the side effects are severe, they can only get worse once the drug is actually on the market.

The lesson is that negative results keep you honest. Reporting them keeps you even more honest. Staying honest is the whole point of science. Physicist Richard Feynman said this about doing science: "The first principle is that you must not fool yourself—and you are the easiest person to fool." Science is a way to not fool ourselves. Admitting and reporting failure is the most important part of that process. Again, Feynman says, "The idea is to try to give all the information to help others to judge the value of your contribution; not just the information that leads to judgment in one particular direction or another." *All* the information includes the negative findings as well.

Now this may all sound so obvious that you are asking yourself, "So how come that's not the way it is?" That's a good question, but not as simple as it seems. Because you can't fix something unless you understand how it's failing, it's worth the time to see what the problem is. Let's have a closer look.

There are two kinds of negative data. We call them Type I and Type II errors. In the interest of not fooling you—or myself—I want to point out, in case you haven't noticed, that I have just pulled a switcheroo on you. I have smudged together the terms *failure, negative results*, and *error*. They are not always the same, although they are obviously in the same family. Why the mix-up? Blame the statisticians (that's what we usually do). Statisticians are especially interested in the particular kind of failure that is due to error, and their methods are very powerful at understanding rates of error, and likelihood of error, and many other things about error. Because many results are "negative" due to a "failure" to reach statistical significance, we use the statistician's terminology for this—error. Thus, we have Type I and Type II errors—which loosely amount to errors of commission and errors of omission.

Type I errors occur when you find something that isn't really there. It is sometimes, and somewhat confusingly, called a *false positive*. That is, your data are wrong because they seem to demonstrate something that is actually not there. Perhaps it is an artifact or a mistaken measurement, but for whatever reason the data seem to show something that is not really the case. Type II errors are when you report that something isn't there when in fact it is, and you missed it. A so-called *false negative*. Both are serious, but both are hard to be sure about. Assuming you do the right analysis with powerful enough statistics of the right sort and that you have produced the data as objectively as possible, using double blinds and all the other ways we try not to fool ourselves, then making a Type I error is going to be hard to catch. They will almost always be found eventually, but usually by some other laboratory which, seeing your excellent looking data, decides to extend that work and therefore relies on your results. This is typically what scientists mean when they say an experiment is replicated, and not that another laboratory has simply tried to re-create your data. Not only is that a boring thing to do, it is, in my opinion, a waste of good time and money. Unless you are really suspicious of some published result, or you have already obtained a very different result at odds with one that was just published, then there is little motivation to simply redo someone else's experiment.

Nonetheless, experiments get replicated because people from other labs use the published results and the methods in their own experiments. If those experiments don't work or they produce anomalous results, then the new experimenters may begin to suspect that the original data being relied upon on was not accurate. Currently there seems to be a lot of complaining in the professional and popular press about this state of affairs. Many published experiments turn out to have unreliable data in them, or results that don't hold up to further scrutiny. I don't see this as a problem, but rather as a normal part of the process of experimental validation. Asking that everything published be completely free of Type I errors would slow down publication to a virtual trickle as labs went through endless repetitions and reanalysis of their data to be sure that it was absolutely correct—and then you could still be fooled. I would rather see things that were done carefully and thoroughly, but not obsessively, get into the literature quickly so that we all have access to it. It is the job of the community to find out what's right and what's wrong with a published result. And it's not always so black and white—things aren't necessarily completely right or completely wrong. Sometimes it takes a new technology or technique to uncover an error made years earlier, so a result may be "right" for some time and then be "wrong" at a later time.

I suspect that a large part of the current public distrust of science is precisely because people are not familiar with this process. The public, for numerous reasons from education to journalism to television, has come to believe that a scientific paper is the pinnacle of the process, the end point where everything is settled. Because they don't recognize it as merely part of a process that will continue to be developed and checked on and validated, they are disappointed when the data don't hold up or have to be revised. But this is a normal part of the process. The key thing is to know at any point how much you can rely on a result and how much it still needs further validation. As pointed out by sociologist Harry Collins, and others, knowing that kind of information is what makes a person an expert in a particular field. I'll come back to all this later in another context. Let's get on with our analysis of errors here.

Type II errors are a completely different story. They carry a special burden, in that it's always harder to spot something that isn't there, right or wrong. These are what are most commonly referred to when we say negative results—that is, failure to find something. They are difficult to publish, and therefore don't often come under the wider scrutiny of the field. Even if they are published, they are rarely given much attention. Who would go to the trouble of replicating an experiment that showed nothing happened?—unless of course you think that something should have happened. This is seen as a problem by many, and here I'm largely in agreement. For the same reasons that clinical trials should publish these kinds of negative results, basic research efforts should also—otherwise you are not following Feynman's dictum to provide all the information to help others to judge the value of your contribution.

Why don't negative results get published? Well some do, albeit in a perverse way. They are the positive results that turn out to be negative when tested by other labs. And although it seems we have no shortage of those, this is not the solution to the problem. Aside from the difficulty of actually recognizing that you have a Type II negative result, there is always the possibility that you didn't get the positive result because you did the experiment wrong, or you made some technical error, or didn't have the best equipment, or any of a thousand things. Perhaps if someone else did the experiment, he or she would have gotten the positive result that really is there. But does publishing the negative result inhibit others from trying the same sort of experiment? Is the second-year graduate student who had just thought of a similar experiment now going to shy away from it since you, the much more advanced postdoc, couldn't get the result? That's not an outcome we want. This is the problem with handling negative results—you can never be sure they are the real thing and not the result of incompetence or some other shortcoming.

It is true that, just like positive results that are later found to be negative, these negative results could later be found to be positive—that is, if others decide to do the experiment you failed at because they think they can make it work. But that's less likely to happen than someone taking your positive finding and testing it by trying to extend it. So there's the added problem that negative results also tend to be long-lived because they aren't tested—they aren't replicated like a positive result would be. Indeed, you never hear anyone complaining about the lack of replication for negative results. But you should. They're no more to be trusted than positive results, and they can be just as important.

This is trickier than you thought five paragraphs ago, eh? Is there a solution to this problem? I think there may be, and we are fortunate because it became available only recently. That solution is Google. Or whoever is smart and adventurous enough to point their search algorithm toward this problem. Now you might think that Google and other search engines are devoted to delivering information that is as true as possible. They aim to get you all the information available about a particular subject. But they don't really do that, because they fail to provide you with the negative information in a usable, dependable format. And this goes against the Feynman dictum. They don't provide you with negative results for almost the same reasons as journals don't publish negative results—there are too many of them, and they're not reliable. A better search may have turned up the positive result.

But Google and other search engines have an advantage over journals. They don't use paper. And they don't need reviewers. They crowdsource. If Google were to set up a site devoted to negative results in science, I believe many scientists would "publish" their negative results there. This would be a citable reference and so could count as a publication. The site itself would best be administered by a neutral scientific society, such as the AAAS, which could convene panels of scientists and journal editors to develop curatorial and review standards. These would be less rigorous, and therefore less time consuming, than reviewing for a journal article. For one, the contributions would be short, presumably a single figure or table of data, and they would mostly concern methodology. Graduate students and postdoctoral fellows might, for example, find reviewing for this site a valuable and educational use of their time.

Remember, the reliability would come not just from the review process, but from the number of times a similar negative result was reported—a number that would be readily available in a Google-like environment. So a Type II negative result would become more and more reliably correct the more often it was reported. It would be reported often because people don't know who is doing experiments that have negative results—because they don't get published. This means they are likely to occur more than once in different labs and would be reported independently at this new website. After some time, as the database builds up, it should be possible to develop analytic tools for quantifying the level of confidence one could have in a particular negative result. It could also become a resource for historians and sociologists of science to investigate the impact of negative results and to better track and understand the process of scientific discovery.

Oh, and by the way, this website should easily be self-supporting. I am sure it would be visited frequently by working scientists—the kind who buy all sorts of expensive scientific supplies and equipment. I can imagine, certainly in my lab, that no one would try an experiment without first checking "NegaData.org" to be sure it hadn't already been tried.

Peter Norvig, Director of Research at Google, told me that he thinks of failure as an aspect of corporate memory. When some new hire comes up with an experiment or idea, some old-timer says, "No, we tried that five years ago and it didn't work." It's not written down anywhere; it's just a matter of corporate memory. This may be positive, saving someone from wasting time and resources on a doomed project. But it may also be that things have changed sufficiently since it was tried so that it may now be a good idea to revisit, with new techniques and fresh perspectives. As Norvig says, the advantage of new, smaller companies may be the *lack* of a corporate memory that stops them from retrying an idea that has transformed from bad to good. But I would say it's best to have both—the information about the negative results *and* the context in which they occurred, so that the current experimenter can judge whether to try it again or try for the same result with a new method. Remember, every negative result had a positive idea behind it, and that idea, even though it didn't work out then, may still be a good one.

A similar dynamic exists in academic research laboratories, especially ones that have been around for a long time. I have written elsewhere that I often advise students to look for new ideas in papers from *Nature* and *Science*—15 years ago. So much has happened since they were published, so many new techniques have been invented that there are questions those authors couldn't even have thought to ask then that we can now. If that's true of the stuff that got published, imagine how much more there could be among the stuff that "failed" then—and we never heard about—but would succeed today.

Now I have to finish this chapter by confronting a very serious problem in the scientific community and one that has gotten a lot of press but little understanding. This is the issue of replication, touched on briefly earlier. This piece could have gone in several different places in this book, and I have been tempted to place it in a few different chapters, verbatim, just to be sure it gets read because I think it's the source of a lot of misunderstanding. For the moment it sits here.

In a paper that has received a great deal of attention since its publication in 2012, two research scientists at the pharmaceutical company Amgen complained that in a review of 53 landmark cancer studies published in high-profile journals, they were able to replicate the results of only 6 of them. This has been characterized as a "dismal" success rate of only 11%. I want you to note the pejorative use of the word *dismal*.

Is it dismal? Are we sure that the success of 11% of landmark, highly innovative studies isn't a bonanza? I wonder if Amgen, looking carefully through its own scientists' data, would find a success rate higher than 11%—or lower? And what did Amgen pay for these 6 brand-new discoveries? Nothing. Not a penny. And if Amgen takes these studies and turns them into cancer drugs that they will charge exorbitant prices for, how much of that money will filter back down to the postdocs and graduate students who did the work? Less than 1%, if any at all. And how much money is Amgen spending to help support this academic research pipeline from which they draw not only results but well-trained future scientists? Not known, but you can bet not much, or they would be crowing about it. (Here I have to make a disclosure—Amgen, alone among the pharma companies to my knowledge, does support a wonderful summer research program, the Amgen Scholars Program, enabling talented undergraduates to work full time in laboratories and attend a special conference each year. My daughter was once an Amgen Scholar, and it changed her life. So perhaps it's a bit ungrateful of me to be using Amgen as a whipping boy here, but finally this all begs the question as to why there aren't more of these programs supported by other Big Pharma companies.)

This so-called dismal success rate has spawned a cottage industry of criticism that is highly charged and lewdly suggestive of science gone wrong, and of a scientific establishment that has developed some kind of rot in its core. Another author, statistician John Ionnaidis, has made a singular reputation for himself by criticizing the apparent low rate of replicability of scientific articles published in major peer-reviewed journals.

What should this really be telling us?

It should tell us that replication is part of the scientific process, not just some post hoc check on it, and that it will therefore include failure—possibly at a very high rate. That publication of a paper in a journal, peer reviewed and all, is not the end of the story and should not therefore be expected to have a 100% correct success rate. Getting to the publication of a paper is a long and tough process that includes mistakes, errors, wrong turns, provisional results, and data done with the best, but nonetheless imperfect, techniques available at the time. Papers and the results in them will all be revised to a greater or lesser, but unpredictable, extent. If you ask for perfection too early in the process, then you will effectively slow the process to a crawl and inhibit the flow of interesting ideas—which can be right or wrong—through the community of researchers. Replication, certainly a crucial part of the scientific process, is nonetheless a slow process. It is often misunderstood by nonscientists—including, I'm afraid to say, statisticians, who do not actually do experiments—as some trivial repetition function, where you do things a bunch of times to make sure it wasn't a fluke. This is not true or even useful replication. And no one would waste their time on it. Exact replication itself is impossible, on purely logical grounds. Indeed, the idea of replication is that it should be done by independent laboratories—in some cases even using different techniques. But even the same laboratory doing repetitions of the work is not going to yield perfect replication. They have already seen the results of the first round of experiments. They are biased. They can't help it. They are humans. Not to mention that maybe they did the first round of experiments in the winter and now they are doing the replications in the summer. Does that matter? I have no idea. And neither does anyone else. In case you think I'm just being extremist in that statement, it has mattered in more than one well-documented case.

In fact, I can't help digressing here to offer a fascinating account of just such a case. In the early 20th century there was a raging debate in neuroscience as to whether the communication between nerve cells was by electrical signaling or through chemicals. Nerve cells communicate through specialized places on their membranes called *synapses* (Greek for "clasping"). These synapses appear as very small spaces between nerve cells, and they may be quite numerous. A nerve cell may have thousands of these connection points with thousands of other nerve cells. That's what makes the brain so complicated—lots of connections. The big question was, how does activity in one nerve cell affect the activity in those cells it is connected with? Physiologists were in favor of bioelectricity, and pharmacologists and biochemists were of course in favor of chemical signaling.

In 1921, Otto Loewi, a German-born pharmacologist working in Austria, dreamed (literally, while asleep) of an experiment that would settle the question. He woke up, scribbled some notes on a piece of paper at his bedside, and went back to sleep. In the morning he could neither read his illegible notes nor remember the dream. For what must have been three agonizing weeks he tried to remember the idea, and then one night he had the dream again. This time, the story has it, he went directly to the lab at 3 AM and set up the experiment.

The details of the experiment are not so important here (and you can look it up easily on Wikipedia), but it involved using the hearts of two frogs, one with its nerve supply still connected and the other just bathed in physiological solution. Nerve input to the heart regulates its rate of beating but is not necessary for the maintenance of a regular beat—the heart does that on its own. Loewi showed that if he stimulated the nerves attached to one heart, it would slow down the heart rate. If he then took the bathing solution from that chamber and added it to the denervated heart, its beating also slowed.

The conclusion was that a chemical had been released from the nerve, and that it was this chemical, and not the direct electrical activity of the stimulated nerve, that affected the heart. For this pioneering work Loewi eventually received a Nobel Prize.

The only problem was that neither he nor anyone else could reliably replicate the results for six years! Why? One reason was that Loewi did the original experiments in the winter. But the frog, being a so-called cold-blooded animal, alters its cardiac physiology seasonally. In all but the coldest months, when Loewi happened to do the original experiments, its nerve input tends to slightly increase firing rate rather than decrease it. Also, the transmitter substance is not broken down as quickly in the cold, so more of it was available for the second heart in the original winter experiments. The season, which you can be sure was not mentioned in the Methods section, made all the difference!

The sociologist Harry Collins points out that in the early days of constructing a then-new special type of laser, the TEA laser, it emerged that the only way to build a TEA laser that worked was to go to some lab that had successfully built one and replicate the process with them. No matter how detailed the instructions, if you had never done it before, you wouldn't be able to do it. This happens every day in science labs of all sorts over far less technical procedures.

It is a commonly held notion that the Methods section of a paper, which describes how the experiments were done, should allow anyone who so desires to replicate the experiment solely from this set of guidelines. This is completely mistaken. It comes from the same bankrupt source as the Scientific Method nonsense. The purpose of the Methods section is to assure other experts in the field that the procedures used were reasonable and accepted methods. And if something unusual was tried, then it is justified and explained more fully. The Methods section allows professional scientists in the particular field to judge that the experiments were done correctly. They are not a manual for doing the experiments. For that you quite often have to call or visit the lab that did the experiments to get all the details, mostly including the unconscious things. Thus it often happens that the methodological instruction "allow the reaction to proceed for about 20–30 minutes" is actually "while waiting for this reaction to proceed, I usually go get some coffee." Which, it turns out, usually takes anywhere from 30 to 40 minutes, and this extra time is actually important.

Replication happens as part of the process and progress of science. Results get published and adopted by other laboratories for their purposes. If the use of these results doesn't hold up, it becomes clear over time that something was wrong with them. Most often it turns up that the results were sort of right, but that things emerge that the original researchers either didn't see or didn't care about. Sometimes those new things are more important than the original results. Sometimes they are just a nice little add-on. Sometimes they show that the initial results were wrong, but they point to a possibility that had not been previously considered and that, approached from a slightly different angle, could still be quite relevant.

All of these scenarios play out every day in research laboratories around the world. And as a result science marches on, failing a bit better every day.

ELEVEN

Philosopher of Failure

Fools give you reasons,
Wise men never try.
— *Some Enchanted Evening*, Rodgers and Hammerstein's
South Pacific, 1949

Once upon a time, actually one day in 1919, the philosopher Karl Popper met up with the psychiatrist Alfred Adler in Vienna. "I reported to him a case which to me did not seem particularly Adlerian, but which he found no difficulty in analyzing in terms of his theory of inferiority feelings, although he had not even seen the child. Slightly shocked, I asked him how he could be so sure. 'Because of my thousand-fold experience,' he replied; whereupon I could not help saying: 'And with this new case, I suppose, your experience has become thousand-and-one-fold.'" This event, when Popper was only an apparently precocious 17-year-old, seemed to have had a lifelong effect on him and led to the development of his most well known, if often misinterpreted, principle of falsifiability as the only dependable marker of a legitimate scientific hypothesis.

Popper arguably has had a greater effect on working scientists than any other modern philosopher of science. And among scientists he reigns as probably the best-known name in the philosophy of science—with Thomas Kuhn running a close second, mostly because of his book *The Structure of Scientific Revolutions*, which gained popularity among nonprofessional audiences and made the phrase "paradigm shift" part of colloquial language. But Kuhn's book and ideas would not be credited by most scientists as changing the way they performed or thought about experiments, whereas many would claim that to be the case for Popper.

In the realm of the curiously contrary, many, perhaps most, philosophers of science now regard Popper's work as seriously flawed and of lesser value (Kuhn, by the way, remains in generally high regard). So the scientists see it one way, and the philosophers another. Not surprising. Without throwing Popper out completely, I'm inclined to go with the philosophers on this one since they are the professionals. Still, whatever you may think of Popper today, his project was an important one and he was successful at framing a question that remains current and troubling.

Popper's original motivation was to answer a simple but dogged question: how can you reliably tell the difference between real science and pseudoscience? How can you know which science stories to trust and which seemingly scientific ideas are nonsense? How do you avoid being taken in by imposters, magicians, con artists, charlatans, and worst of all, well-meaning, devoted practitioners of marginal pseudoscience? If this seems at first trivial, I might point out that this question has not yet been answered satisfactorily, even in our technologically advanced Western cultures, let alone in cultures that remain dominated by so-called folk wisdom and various magical "explanations" about how things work. Based on recent history—from the fiasco over vaccines and autism to the conspiracy theories about the causes of AIDS, it seems that otherwise intelligent people cannot reliably tell the difference between scientifically valid explanations and pseudoscientific malarkey. Part of the problem, as Popper realized, is that legitimate science is sometimes wrong and pseudoscience occasionally stumbles onto something true.

Every year millions of people are infected by influenza, a disease that can still kill thousands of people worldwide. The name *influenza* comes from the Italian for "influence" because it was believed that the disease was due to unseen celestial influences, as predicted by astrologers. At the same time, we are told that the oceanic tides are caused by unseen celestial influences as described by Newtonian theories of gravity. The latter of course is true and scientific, while influenza is actually caused by . . . unseen microbial organisms.

We know this now, but you can see the 17th-century difficulty in deciding that one invisible lunar influence is a false reason, and the other equally imperceptible lunar influence is the real thing.

In our own day we have Scientology, a made-up, silly word intended to seem more scientific. We have intelligent design. We still have astrology. Think these are just easy targets? How about GMO foods? Nuclear energy? Natural products? Alternative medical treatments that often forestall life-saving standard medical practices? (Here I think of the brilliant Steve Jobs eschewing surgical treatment for his early stage and curable form of pancreatic cancer, electing various diets and other alternative treatments until the disease had reached a fatal stage.) Educated, intelligent people have strong opinions and beliefs in these completely nonscientific concepts or practices. They are often the basis of critical policy decisions that affect many millions of people and have serious economic consequences. They are not science or scientifically based, but they present themselves as science.

Back to early 20th-century Vienna, where many wild and revolutionary ideas were being avidly discussed and debated in the city's famous coffeehouses. Two ideas were of particular interest to Popper: Einstein's relativity and Freud's psychoanalysis. Einstein's theory, in which mass and energy were no longer distinguishable and time was malleable, seemed every bit as crazy as Freud's depiction of the human mind as an unimaginable seething organ of infantile jeal-ousies, neuroses, and hysterias ready to break through the barely controlling ego at any moment. But then Sir Arthur Eddington reported on his observations of starlight passing close to the sun during the solar eclipse of May 1919, con-firming one of Einstein's most radical predictions—that gravity curved space and would alter the path of photons. At the same time, Popper saw that Freud found continual confirmation of his theories no matter what the observation. As with the Adler anecdote, there seemed nothing that could refute the psychoanalytical structure that Freud had created. Every new occurrence, even if apparently contrary to the currently accepted dogma, was not only twisted to fit it, but then served as yet further confirmation of the theory. There were never any refutations, and indeed no possibility of there being one. Popper concluded that this could not be the basis for scientific thinking, not because you couldn't prove it, but because you could *not* disprove it.

Popper's solution was a tremendous insight. He recognized that this difficulty of showing that a discipline was truly sci-entific stemmed from the fact that in any system of belief you could always prove things right, always find an explanation that would account for the most anomalous of findings. Sci-ence couldn't claim to be science because it was right. That would make it just like any other belief system. What made something a scientific statement was that it was possible to prove it wrong. Science according to Popper made risky predictions which could be empirically tested and which, if shown to be contradictory to the theory, would require that the theory be abandoned. Thus a legitimate hypothesis was one that could be shown to be incorrect by empirical scientific experimentation. Science was dependent on failure, or at least its possibility.

This does not mean that a hypothesis has to be wrong, but it must have the possibility of being shown to be wrong. Any scientific hypothesis is then only provisionally correct, because at any time an experiment could be performed and the results could show it to be wrong. Of course the more such experiments are performed and confirm the hypothesis, the more likely it is to be correct—but still, one negative result could theoretically overthrow the whole thing.

This is most commonly presented as the famous case of the black swan. For a long time, all the swans that anybody saw (in Europe, of course) were white. One might consequently form the hypothesis that all swans are white. Then black swans were discovered in Australia. Therefore, the "all swans are white" hypothesis is overthrown. The point is that the hypothesis allowed for a negative test—the observation of a black, or any nonwhite, swan would overthrow the hypothesis. This characteristic makes the hypothesis a scientific statement. A nonscientific statement might be that all swans come from heaven. This can neither be proved nor falsified because one can always say that the ultimate if unobservable source of a swan, no matter how many times you see an egg hatch, is heaven. Because there is no way to falsify this theory—heaven, by definition, can do anything it pleases—it is not a scientific statement.

Popper's elegant statement about scientific hypotheses is based on fallibility. Science is trustworthy precisely because it can fail. A hypothesis is to be trusted only if it can be falsified. To the extent that scientists actually use hypotheses (see chapter 8, "The Scientific Method of Failure"), this is good advice. Framing a hypothesis is the art of suggesting experiments that could fail easily and which would then negate the hypothesis. If a hypothesis is not clearly falsifiable, then it is not an acceptable hypothesis.

So what's the problem with this?

First, it doesn't always provide a clear result. According to the Popperian prescription, the detection of aberrations in the orbit of Neptune should have provided a reason to throw out Newton's celestial mechanics; instead, it led to the prediction and discovery of the then-invisible planet Uranus, reaffirming Newton's laws in the process. On the other hand, the case of Mercury's departures from the predicted orbit would also be a falsification of Newtonian ideas of gravity—and in this case it was. The Mercurial aberrations could be accounted for only by the relativistic model of gravity. So Popperian falsifiability is not perfect—it fails now and again to correctly identify a useful failure. In the end you are still stuck with making a judgment about how important the failure is compared to the theory. And some scientific endeavors simply don't allow for experimentation of the sort Popper would demand. Evolution is one example of a science where those kinds of experiments are hard to construct.

Another difficulty is that most theories do not stand alone but are embedded in a fabric of ideas and other theories that they refer to or depend upon. When an experiment provides a contrary result, it is often hard to determine where the fault actually lies—and you don't necessarily want to throw everything out because of a minor failure in one part. More sophisticated arguments attacking the very logic of the proposition have also been made—and answered. For example, Kuhn claims that virtually all theories are falsified in some part at some time. I agree with this—science is a series of provisional findings that iteratively moves us closer and closer to a truth that may never be fully attained. But the provisional iterations are valuable even though they are ultimately wrong. They shouldn't be discarded, which is what a strict adherence to Popperian rules would have us do.

In a bit of an ironic twist, while Popper elevates the importance of failure and falsifiability, if taken literally his litmus test would exclude failure from the process of scientific advancement.

Popper was right that a hallmark of true science is failure. It may not rise to the level of proof; in fact, there may not be any proof that some activity satisfies the credentials for real science. I doubt it will ever be that easy—or that stable an idea. What we can say is that failure is an indelible and continuing part of any scientific activity, and if it is missing, then the likelihood that such an activity can be regarded as science is severely diminished. One should be more suspicious of extravagant success than regular failure. Popper had the right idea, but it fails, perhaps to his credit, to fulfill the final purpose he had in mind for it.

TWELVE

Funding Failure

In the long run failure was the only thing that worked predictably.
—Joseph Heller

W orld War II was largely fought—and won—by science and technology. From aviation to rocketry, radar, sonar, cryptography, and of course the atomic bomb, science dominated the course of this war more than any previous conflict. By the end of the war America was in possession of the most developed scientific infrastructure the planet had ever seen. What were we to do with it? What role would this advanced wartime science play in peacetime? President Franklin Roosevelt, who during the war had appointed Vannevar Bush (no relation to the current political dynasty) to be what was unofficially the first presidential science advisor, now tasked him with the challenge of transitioning scientific research from defense to peacetime initiatives.

The directive was expansive in its scope, giving Bush leeway to consider everything from medicine to security to education. The result was the now-famous report *Science, the Endless Frontier*. With this encompassing script Bush effectively set up the structure for making scientific research largely the responsibility of the government, even when it was not defense related. The value of science to society—the cures, jobs, economic development, education, and general well-being that it could bring to a modern life—was to be primarily the responsibility of the government. Only the government could direct an effort comparable to that mounted in the cause of wartime victory that would allow science to flourish in the service of a wider public good.

Although agencies like the modern-day National Institutes of Health (NIH) existed and were funded by Congress, it was more a Public Health Service (its actual name, originally) than a serious research organization. Similarly, the modern National Science Foundation was created in 1950, alongside several other then-subsidiary departments also dispensing federal funds for scientific research. But it was through Bush's efforts that these agencies were expanded to their current status as the prime supporters of research of all sorts in the United States. His vision has become the worldwide model for the pursuit of science. Although Bush wanted a plan that provided even more independence for science from politics, and a more integrated program for the physical and health sciences, his policies significantly shifted the postwar attitude toward science as one of societal responsibility and ownership. As with the war effort, there would now be an equivalent effort to use science for the betterment of humankind. And it would take government to manage such an immense effort.

In so doing, we tacitly recognize that paying for science collectively, with tax dollars, is a worthwhile investment because society as a whole, as well as many of its members individually, reaps the benefits. One could make the case that the standard of living in scientific societies is better by far than in any societies where science is or was not an integral part of the social organization. A second reason that most science is funded by the government is that it has become too expensive to afford except by spreading the cost over society. It has become expensive in two different ways. The equipment and related costs for performing many kinds of experiments has become pricier, and, more importantly, the number of scientists has increased dramatically. We now have more people doing science than ever before. Indeed, it is estimated that there are more working scientists alive today than the total number of scientists who have existed since Galileo (around 1600). Accordingly science appropriations have risen over the years, although as a portion of the federal budget and a percentage of the GDP, the increases have leveled off and even gone down in recent years.

You might ask why industry doesn't pick up some of the tab here, because it can certainly afford it. Of course some industries do participate in their own research and development programs, but none supports the kind of open-ended research that is necessary to maintain the overall health of the enterprise. That work would be too risky and adventurous for the captains of industry to commit to. Better to leave that to the academics and then use that government-sponsored research, almost entirely free of charge, by the way, to develop profitable new drugs and devices. Perhaps that sounds like sour grapes, but I have to admit that this model has worked, and worked well for almost everyone for quite a long time.

For all the general good created when science is so predominantly funded by the government, there are going to be pernicious effects, and it is important to keep them in mind. For one, science now becomes a matter of policy, directed by what the government, in the guise of review committees or advisory boards or congressional mandates, wants to know, which may not be what nature is offering up. There are agendas and initiatives, challenges to be met, diseases to be conquered, and so on. Of course, spending all that money should be as well planned as possible, but it shouldn't be any more planned than absolutely necessary. Overestimating how well we can chart science is a fatal hubris that seems to strike even well-meaning people, including practicing scientists, when they are placed in an advisory or administrative capacity. Finding this set point between responsible management and counterproductive control is not trivial. It is in many ways, even more than actual dollars, at the heart of the current crises in science funding.

Whether it is because money in general, and specifically money for research, is tighter now, or because the citizens who pay for it have become fragmented in their support for science, or other reasons that sociologists can and will puzzle over, the fact is that the landscape for science funding has changed. And it has not, for the most part, changed for the better. With all the squabbling for dollars and programs, it may seem impossible to pick out a single cause for the financial malaise currently afflicting science, but I would suggest that underlying it all, there is one major change that has had the most injurious effect on the practice and culture of science: we don't fund failure as well as we used to.

In the 19th century, men of means were the only ones who could afford to be scientists. The word *scientist* was invented to describe them (coined by Cambridge polymath William Whewell in 1833). Most scientific progress was then the result of personal wealth allowing time for dabbling and experimenting—and failing. After all, it was your money, so if you wanted to piss it away on some grand theory or another or some quest for understanding mysterious phenomena, then that was your right. But it meant that a landed gentleman like Charles Darwin could afford to finance his 5-year journey of exploration on the *Beagle* and the expenses associated with collecting, shipping, and maintaining an extensive collection of specimens—and then take more than 20 years to cogitate over his data and develop his theory.

Likewise Gregor Mendel is famously characterized as an obscure monk in an out-of-the-way abbey in the smallish town of Brno (then Austrian, now Czech) who had a green thumb and played around with pea plants while discovering the gene. In fact, Mendel had the protection and generous support of the abbot for his experiments, which hardly amounted to tinkering in the garden. Mendel is estimated to have raised at least 29,000 pea plants, the result of many tens of complex crosses as he sorted out seven different genetic strains over several generations. His was a major undertaking, and it was widely reported in the scientific literature of the day, although later somehow forgotten (this, by the way, is not all that unusual a story in science). The point is that the supportive abbot, and Mendel's position in the abbey, gave him the time to devote to these painstaking experiments—over more than 7 years.

We underestimate the importance of patience in science. Patience comes along with having the option to fail, from having the opportunity to pursue promising ideas that may ultimately turn out to be wrong, or not completely right. That kind of funding support has been disappearing at an alarming rate over the past decades.

Funding levels for scientific activity have become so constrained and so competitive that researchers have learned to propose only projects that they are sure will work in their grant applications. It is not uncommon for half the experiments to be done before the application even gets submitted—just so researchers can later claim success in order to be assured of future funding. There must be enough "preliminary data" to virtually ensure the success of the proposed experiments. Sydney Brenner, Nobel laureate and maverick biologist, only half-jokingly describes the standard NIH grant as having two parts: the first part proposes the experiments you've already done, and the second part proposes experiments you will never do.

Failure is being driven out of science by the lack of funding and the resultant increase in competiveness. Has this been good for science? We have to look carefully at this one—the answer may not be as clear as it was in education. There is a case to be made for reducing waste, for spending our limited money and resources more thoughtfully, for getting the biggest bang out of the buck. On the other hand, it is hardly an efficient way to spend resources if what you are getting is simply incremental advancement in areas of virtual certainty with ever-decreasing chances of making unexpected breakthroughs.

The NIH has an obscure system for identifying different types of grants. One of the categories is called a "High

Risk/High Impact" proposal. At the National Science Foundation (NSF) this is often called "transformative research," presumably after Thomas Kuhn's famous dictum about paradigm-changing research. The idea is the same: the research proposed is risky and uncertain, and it has a high likelihood of failure—but if it works, the rewards will be great and possibly "game changing." In 2013, a total of 78 such grants were awarded out of approximately 5000 grants. That's around 1.5%. What I wonder is, why do we need such a category? What does this make all the rest of the research we are funding—predictable, pedestrian, likely to succeed, but of minimal importance and impact? Is that where we want most (98.5%) of the money to go? Haven't we invented this high-risk category in response to our timidity about failure? It's a tacit recognition that we can't just have guaranteed research, but at least we're going to restrict the failures to a smaller, identified arena. But there is no smaller, identifiable, and separate place in science where failure can be contained. Failure is, or should be, embedded in all of it. Fear of failure is an institutional weakness, but it will quickly spread to individual laboratories because of misguided policies. Especially dangerous are the ones that sound good, like *High Risk/High Impact*.

Science at its best is a grand marketplace of ideas. Like any marketplace it can be regulated only delicately without interfering with its greatest strength—the possibility of unexpected outcomes from unexpected sources interacting and intersecting freely with one another. This description, already messy, doesn't even begin to catch the real messiness of the process. And it does not capture the high rate of failure that is to be expected from a messy process (see entropy, Chapter 3).

Unfortunately, there is no real way to gauge the failures and therefore no way to effectively set or manipulate the rate of failures. This is, in large part, I think, because of the inescapable and implacable element of time. What may look like a failure today may be destined to be a success at some later time when new data become available and unexpectedly a previously concealed value appears. There are piles of these stories in the history of science. The laser was considered to be an impossible device by most of the great minds of physics in the 1950s. Charles Townes, then a 30-year-old physicist at Columbia University and Bell Labs, was advised on more than one occasion to drop his investigations into coherent light and to get serious and stop wasting time and money on this quixotic effort. No company would have invested a dollar in this risky, sure to fail, research in the 1950s; but what industry could operate without lasers today? Townes shared a Nobel Prize for this work in 1964. But that took 10 years, and the scientific and commercial development of the laser has taken far longer—indeed, it is probably still not complete. So how could this work have been judged a success or a failure at its inception?

Like any marketplace, science is easily undermined by overzealous central planning. The best modern example is probably the disastrous consequences of the Soviet Union's state-sanctioned support for Trofim Lysenko, who championed a kind of Lamarckian genetics. The attraction of Lamarck's genetics was that gains made by one generation could be passed on to its offspring. Hard work and self-improvement would be passed on to your children—a near perfect Marxist biological principle. Unfortunately, it's not correct, and the pursuit of this false genetics without competition from other ideas led to decades of crop failures (at the same time as the application of Mendelian genetics to agriculture was producing record crops in the United States and Western Europe). There are many reasons for the collapse of the Soviet Union, but is it going too far to suggest that the starvation of the population through the use of genetic methods that were not scientifically valid but rather state sanctioned was a contributing effect? Not all cases of state-directed science will go this far off the rails, but they will surely all have some political motivation behind them. At this writing, one funding bill for scientific research in the United States includes specific, and drastic, cuts for research in the social sciences—because their results are rarely favorable to that particular political party's economic viewpoints.

By the way, there is an interesting update to this story that shows the value of failure. Lamarckian genetics may be an essentially wrong and a largely failed theory of inheritance, especially as compared to Mendelian genetics and Darwinian evolution. But it is currently making something of a comeback in the field of epigenetics, where certain behavioral traits or environmental factors are able to alter the adult genome and be passed on to offspring. But it is only in the context of a free exchange of ideas that Lamarckian genetics is able to make a reappearance and find its appropriate place in our wider understanding of genetics. In its isolated form it is just a mistake—which is less, much less, than a failure.

An oft-proposed alternative is that we just embrace serendipity, which is how it seems the popular press thinks most scientific discoveries come about anyway. Rather than go off on a long tangent here about why I think that is wrong (see chapter 3), let's take it, for discussion's sake, to simply mean "an unexpected discovery." Serendipity has become a cornerstone in the argument for funding so-called basic, or fundamental, research—that is, research whose goal is to increase knowledge without having a specific application in mind. The reasoning goes that since we are not smart enough to predict where the next advance will come from, the only sensible strategy is to simply fund the most interesting research into fundamental questions and gratefully reap the benefits of the ones that work.

But that argument never seems to wash. Everyone likes the idea of serendipity in science—until it comes to funding

it. Then anything that says "We're going to try this interesting but novel idea out and hope to get lucky" is sure to
find its way on to the rejection pile. Indeed, one of the worst
critiques you can get from a review panel is that your proposal is just "curiosity-driven" research. I know that seems
ridiculous, but NIH and NSF require that all research be
"hypothesis-driven" and that it be proposed in the form
of multiple hypotheses to be tested. Screw that curiosity
thing: that's for kids and creative types. Hard to imagine
a branch of government devoted to science and filled with
supposedly smart people coming up with a provision like
that—and worse, sticking to it for the last 50 years.

The problem of having too little adventure in science—that
is, not enough leeway for unusual ideas, keeping on the
straight and narrow, if you will—was recently highlighted by
a group of prominent English scientists. In an open letter to
the *Guardian* (UK) newspaper of March 18, 2014, titled "We
Need More Scientific Mavericks," 30 of Britain's top scientists
claimed that the great scientific advances of the 20th century
and earlier were accomplished because there was support
for people who thought differently and were not required to
prove immediate value or use for their research. They went
on to note that this unfettered research had resulted in such
advances as the transistor, the laser, electronics and telecom-
munications, nuclear power, biotechnology, and medical
diagnostics—a short and partial list. A similar letter, with

the addition of some Nobel laureate signatures, appeared in the *Daily Telegraph* under the title "Nobel Winners Say Scientific Discovery Virtually Impossible Due to Funding Bureaucracy." Donald Braben, an earth scientist at University College London, recently published a book describing 500 important discoveries that arose from maverick, curiosity-driven research.

While working on this chapter I received a research grant funded by the National Institutes of Health. It underwent the review process common for NIH applications and was awarded a high enough score to rank in the 8th percentile of applications reviewed in the same period. By current standards that is good enough to make the "payline" (a term that sounds uncomfortably like the threshold for paying off a wager on a sports score or some other game of chance), and therefore my laboratory will receive funding to support our research over the next five years. I should be quite happy. And I am certainly happier than those whose scores fell below the payline and who will not get funding. But I wrote several proposals to get that one funded, turning the process into a kind of lottery—the more you play, the better your chances (which is actually not true; it's just the only option).

This all-important payline varies among the several institutes—the National Institutes of Health (it's a plural) stands for an umbrella administration of 27 separate institutes devoted to various biomedical concerns. There is the

Eye Institute, the Cancer Institute, the Institute of Infectious Diseases, and so on—you can look them all up easily enough. Each institute has its own budget, sets its priorities, and funds grant applications at different percentile levels. This time I'm fortunate because my institute is funding at the relatively high level of around the 17th percentile. Some institutes fund at below the 10th percentile, and once I submitted a grant to an institute that turned out to be funding at the 2nd percentile! Needless to say, that went nowhere.

Is it really possible that only between 2% and 20% of the grants that are submitted are worth funding? Remember that in order to submit a grant you must have at least a PhD or MD and be employed at a recognized university or research institute. So the 25,000 or so grant applications submitted annually to NIH (and 90,000 to NSF) do not follow a normal distribution; they are already a highly selected assemblage. Imagine the really good science being wasted, dying in someone's bottom drawer as an unfunded proposal, a clever idea or a new concept that will never even be tried. Is this a sensible use of this unsurpassed resource of well-trained sophisticated scientists that has never been available anywhere in the world or at any time in the past?

In the biological sciences the most successful, and perhaps instructive, example of funding failure is in the area of cancer. In 1971 then-president Richard Nixon declared a War on Cancer, because wars appear to be the metaphor for

Americans going all out for something. Be that as it may, $125 billion has gone into cancer research since then, and funding has recently leveled out at around $5 billion per year for the last 4 to 5 years. The result: over the same 42-year period, some 16 million people have died from cancer, and it is now the leading cause of death in the United States. Sounds bad, but in fact we have cured or developed treatments for many previously fatal cancers and prevented an unknowable number of cases just by establishing the importance of environmental factors (asbestos, smoking, sunshine, etc.).

And what about all the ancillary benefits that weren't in the original prediction—vaccines, improved drug delivery methods, sophisticated understanding of cell development and aging, new methods in experimental genetics, discoveries that tell us how genes are regulated—all genes; not just cancer genes—and a host of other goodies that never get counted as resulting from the "war" on cancer? We have learned unexpected things about biology at every level of organization—from biochemical reactions inside cells to regulatory systems in whole animals and people, and the above-mentioned environmental effects on health. Is anybody tallying all this up? I'd venture to say that, all in all, this cancer war has given us more return for the dollars spent than any military war, certainly any recent war.

Nota bene: Most of this research focused on finding a cure for cancer has been marked by failure. Hundreds, probably thousands, of researchers have ventured down one path and another, often found out some very interesting things, but failed to cure cancer. And for all of that "failure" this process worked so well for so many years because the Cancer Institute, like most of NIH, had traditionally funded grants at the 25th to 30th percentile. But by 2011 funding had dropped to the 14th percentile, and by 2013 it was at the 7th percentile—only 7% of the grants submitted were funded to pursue research into the leading cause of death in America. Is that what we want? Isn't there something very obviously wrong with this picture?

Maybe the better question is, is there a fix? I think there are several possible fixes, but they will all take some courage.

The easiest suggestion to make is that funding levels for science, of all sorts, should be increased. This is of course easy to say—and easy to criticize as just "throwing money at the problem." But the thing is that money is often thrown at problems, and it often works—at least well enough. Our main defense strategy is to throw money at the military for all sorts of projects, many of dubious value. In one of the more ironic cases of money throwing, in 2008 the government threw a great deal of money at monetary institutions like banks and investment houses to save them from a failure of their own making and stave off a depression. And that seemed to have worked. So why is it okay sometimes but not others? Why is increased spending on defense a sensible policy while spending more on education, research, and social problems is considered wasteful? In the case of the banks, the excuse we were given was that the financial institutions were too big to fail. Don't we think that the scientific research infrastructure that has given us innumerable cures and technological advances is too precious to fail?

Perhaps one problem is that the choices for funding science or education are so varied and it is difficult to know which of many good ideas and proposals are likely to succeed. In science you may have a genuine plurality of choices that are hard to decide among. How best to pursue cancer research: through cellular mechanism studies, immunological studies, epidemiological studies, clinical studies? And even within each of those areas there are dozens of strategic options. But this is also the great strength of science—there are many options, many ideas, many *good* ideas. We don't want that to be restricted. We don't want to shrink the pool of good ideas, even if we can't decide which ones are going to prove most successful. Indeed, it is precisely because we can't predict outcomes with much certainty that we need to maintain as many approaches as we can. This is known as hedging your bets, and it is a very good strategy practiced by many investors and gamblers—to the extent that those are different populations. I suspect it's a good strategy in science as well.

All this notwithstanding, the real problem may not simply be insufficient funding. As with many similar situations it is the distribution, rather than the absolute amount, that is the most pressing problem. Distribution may also be easier to fix than lobbying an already overburdened system for additional allocations. So, yes, more money would be good, but looking at other options may not only offer alternative solutions; it may also reveal some unexpected perspectives on funding science.

Let's have a look into the current methods for government funding of science, by which we really mean government funding of ignorance (it's the unknown we're after). Government oversight is of course proper since science is paid for from the common coffers. However, the government needs to be careful about defining the scientific agenda too carefully. Instituting programs with definitive goals and focused questions, while tempting for policymakers to consider as being efficient, runs the risk of defining ignorance too narrowly. There has to be room for unusual ideas that don't neatly fit a programmatic goal. We must guard against having too little adventure in science. Easily said, but we have to recognize that it is difficult to build these sorts of risky incentives into a bureaucratic funding structure that is also responsible for monitoring the results of its decisions.

How, then, does the money get distributed? How are the decisions made? Here's a bit of an insider view of the grant process we use in the United States today, specifically at the NIH, where I have had the most direct experience, as reviewer, proposer, and occasional recipient. You hear a lot of moaning and whining from scientists about grants and the whole process—so here's a view with some details you probably don't know. But without knowing them, how are you supposed to know what to think about all this? Remember, taxpayer, this is your money we're talking about.

In the NIH review process, a grant proposal is supposed to be scored on several categories: Significance, Investigator, Innovation, Approach, and Environment. Although all are supposedly given equal weight in the final scoring, Approach and Innovation are the scores that typically make or break the proposal. Not that the other categories—the significance of the work or the scientists who are going to be doing it or where it is going to be performed—are unimportant, but they are usually givens. Of course it's significant, at least to you; why else would you propose it? And grant applications are submitted from PhD- or MD-holding faculty members at major universities, so they virtually always meet the Environment standards. It may happen very rarely, but I have never heard a proposal seriously criticized, let alone turned down, for having insufficient facilities or an incapable applicant.

Approach is the main part of the grant application because it details what experiments are going to be performed and how the data will be analyzed and what the results will mean—hopefully. One of the required elements of the Approach is a consideration by the grant writer of the potential problems and pitfalls, including an often cursory description of what will be done to avoid these or what will be done in the event that they happen. Although this all seems reasonable enough, this is after all the "failure" part of the proposal, it is in practice a monstrous error. Whether intended or not, it forces the applicant to write the proposal from the perspective of success. It becomes a sales pitch. The only experiments proposed are the ones likely to work. In the very unlikely event that they don't, then we can propose some workarounds. But this isn't real science—not the kind that makes really novel discoveries and opens up new inquiries. If all or most of what you propose works then most likely what you will have at the end of five years is sufficient material to use as preliminary data in your next grant application, so that you can assure the reviewers that there is a high probability things will work for another five years. And so on goes the cycle through another career in science. Of course everyone knows the game, but this makes the grant proposal just that much more of a shallow sales pitch.

Even worse, these sales pitches create unrealistic expectations about what can be delivered and how rapidly it will all come to pass—that, after all, is what a successful sales pitch does. Competition among the sales pitches—grant proposals, if you like—forces the promises to become ever more inflated. Then, when the science doesn't deliver quickly enough, politicians call for cuts in funding or a shift to translational research. The Human Genome Project is perhaps the most obvious case of this malignant cycle of unrealistic promises followed by unwarranted cuts. It promised a revolution in medicine, and while it finally delivered on providing the actual sequence of the human genome, it has not led to any identifiable clinical advances. Ironically, it is one of the most important tools of biologists doing basic research—and one day those research programs will impact the clinic, if we don't starve them to death first. But we must apply patience and we must recognize that basic and translational research share the same pipeline. You can't shift from one to the other as if there were two spigots.

Innovation as a grant category is yet more dangerous. Here the reviewer must undertake the ridiculous task of quantifying how innovative something is with a number (score) between 1 and 9. He or she must then write a few terse lines about what is "innovative" in the proposal. I suppose this is intended to stand in for the perhaps too childish idea of curiosity, or the difficult-to-evaluate qualities of thoughtful ignorance and creativity. But it doesn't. *Innovative* now has come to mean "novel" or "new." Dazzle me a bit with what's never been done before. But innovative for the sake of novelty is not the same as curiosity. Innovative often regresses to "technically sophisticated"—using some new technology or equipment. It can, to be fair, mean taking a novel approach to a longstanding question. But in my experience as both reviewer and applicant, that is rarely the case. Innovative is therefore perilous territory for the grant writer. You don't want to be criticized for being dull and boring, doing incremental research. But if it's *too* innovative, then feasibility goes down a point or two and that can sink the application. Although a thoughtful reviewer can use this category to score points for a creative proposal, it's just as often the category where a more conservative reviewer can shoot down the application as frivolous.

All of this monkey business, this grantsmanship and arbitrariness, has arisen from the attempt to standardize grant reviewing, to make it fit into five scorable categories. Having rules for the game would seem on the surface to be a good idea—leveling the field and giving everyone an equal chance. The NIH website explains "How Reviewers Score Applications," insisting that the overall score for a grant proposal is "more than the sum of its parts" and is an "integrated gestalt." But anyone who has been on a review panel (called Study Sections, with the unfortunate acronym SS) or had their grant application reviewed by one will tell you that is just mumbo jumbo. It's scores all the way down. And the trouble is that scores don't do well at rating curiosity, imagination, failure, and uncertainty. There is no Maverick score. And because they don't do well at evaluating those categories, they are precisely the categories that get dropped out of the process. If this is the cost of leveling the playing field (which none of this does effectively anyway), then it's a very worthy, but very bad, idea. That happens sometimes.

To get all this working again, there must be a change in perspective. First, we don't fund success; we fund failure or the potential for failure. If we want science to tell us about things we don't yet know—its only reasonable purpose it seems to me—then we are going to fund a lot of attempts that don't succeed, at least not right away or in precisely the way the goals had been set. This means that the way we judge what to fund cannot be a top-down decision, since we are then asking a few people to make decisions about failure—and that's asking too much of too few.

We are caught in a bind. We try to make our decisions based on the likelihood of success, but we know from experience that we are poor at making that judgment and there are serious losses that come from failing to recognize a good failure. But judging potential failures is even more difficult. How do you tell a good failure from a bad one, a useful one from a waste of time, an informative generative failure from a dead end?

In my opinion there are two possible solutions. The first is in many ways the best one, but I think it may be too radical to be adopted. This would be a system in which funding decisions are made by random selection after a fairly minimal filtering or triage review. Donald Gillies (University College London), using an unpublished essay by the late Sir James

Black (Nobel Prize–winning discoverer of beta blockers and cimetidine—Tagamet, to many of us) suggests random funding of applications after a cursory screening for proper scope and competent personnel (i.e., filtering out the cranks). Many others have done extensive work to support this revolutionary but growing view, among them a bright graduate student at Cambridge named Shahar Avin (with whom I was able to discuss this over many warmish beers), and several groups in Australia, the United Kingdom, and the United States.

Given the number of hours that are currently devoted to writing and reviewing grant proposals, we might realize significant savings by using a simple lottery method. Nicholas Graves, a health economist and professor at Queensland University of Technology, estimated that in 2012 Australian researchers put in the equivalent of five centuries of work on grant writing. And since only 20% of the proposals were funded, that means four centuries of work went more or less wasted. Note the units here: centuries! Reducing this cost alone would be sufficient reason to adopt a random system if it could be shown that the results were not appreciably different from the current review process.

Of course you can imagine the hue and cry that would rise from legislative bodies entrusted with the public bankroll. And I must admit that, convincing as the argument may be, I don't really like the idea that science research could or should be supported by random processes. It seems so . . .

unscientific. It is also an argument from failure of the bad sort: the current system is cumbersome and expensive and failing to produce the desired results, so let's just throw it out and throw darts, because how much worse could it be? Are we really no better than Rabelais's satirically drawn Judge Bridlegoose, who spends hours reading and pondering a mountain of documents before deciding by rolling dice, a method he claimed to be as reliable as any other? The random strategy emerges as reasonable mostly I think because the current system has become so warped. But perhaps it could be unknotted and returned to its historically more productive state.

To do that we need the second possible solution, and the only one that is to my mind realizable. I suggest that we go back to the market model of science funding. Not the least because it worked. Competition among proposals should be based on the merit of their science and the creativity of their approach, on the quality of their curiosity. While eventually this will require more money in the pot, a great deal can be accomplished immediately and even at current spending levels. There are two conditions that must be satisfied for the market model to work. First, there has to be a reasonable expectation on the part of an applicant for obtaining funding; otherwise, he or she would have to continue to write sales pitches instead of grant proposals. Second, there has to be a margin for failure that is large enough to permit creative thought, but not so large that we dispense with judgment.

In biomedical funding I would suggest that this failure margin should be around 30%—it was the number that built US biomedical science into the world-leading powerhouse that it had been from the postwar years to the final decade of the 20th century. The current crisis of decline can be traced from the lowered paylines caused by no-growth budgets that failed even to keep pace with inflation and, I believe even more importantly, top-down earmarking for certain kinds of translational science projects. If all of those top-down earmarks were ended and the money put back into individual grants undergoing peer review for scientific merit, then the paylines would be up to within a few percent of the historical levels. The most important, and the simplest, correction would be to simply reverse the past 15 years of failed policies and revert to the original model of peer review by expert panels, with fewer directives from the administrative layer.

This would have two immediate and salient effects, both of which I believe are critical to a successful funding policy. For the application part of the process, the higher paylines would restore rationality to grant writing by creating a reasonable expectation of funding for honestly presented proposals. On the evaluation side, review panels are better at including failure in assessing value than are earmarked programs, which set out goals and with them expectations that are often unrealistic—and never take failures into account. The peer review process is not perfect, but it can be worked on and improved. Earmarks are law. They are resistant to change once in place. They are not the way to run science, which is characterized, if nothing else, by constant change and revision.

I don't know what the numbers would be in physics or chemistry or mathematics or psychology or various environmental studies, but I bet they would be around the same. Actually, I don't even know that 30% is the correct number in biology today. But these numbers could be determined, and determined I believe to a sufficiently accurate value. Using historical data and good mathematical models, we could work out the minimum amount of research that should be funded to ensure continued success and rational competiveness. This would also have to include estimates of what it would take to maintain science as a desirable career path that attracts young and talented people—all of whom could instead go into finance and make a killing. That would take some work, but again the data and the mathematical tools are there to do the job. This is one area where we can apply quantitative measures and years of data to get a useful value. At the worst we would obtain a good estimate for a starting value and then adjustments could be made as new data becomes available. What a grand experiment!

So the opening strategy is to raise paylines by simply removing top-down earmarks for specific programs and letting the marketplace of ideas and reviews determine the best place to invest our scientific budget. Eventually, though, spending must also increase. Given the number of studies showing that investment in scientific research creates the greatest return of any government spending program, I can see no reasonable argument against throwing some money at the problem. Of course throwing money should be done as wisely as we can. The problem is that we are not very realistic about our level of wisdom.

We continue to have the most important resources at hand—eager students and postdocs who are well trained and are ready to step up to the job. The next generation is here and ready to work—ready to fail and to fail better. It pains me to see that when people suggest cutting science budgets, their first target is commonly graduate education. Yes, fewer graduate students will take the pressure off of grants since there will be fewer people competing for the same dollars. And whom will we wind up excluding this way? Does anyone really have any idea how to decide who will become the great scientists and who should be shunted off to other pursuits? Why restrict the single greatest resource we have—graduate students and postdoctoral fellows with new ideas? After all, they deserve a decent chance to show us how brilliantly they can fail.

THIRTEEN

Pharma Failure

The Second Law of Pharmacology: The specificity of a drug decreases with
the time it is on the market.
—Unknown

I love Big Pharma. They are the biggest and best failers in
science. They fail a lot and they fail reliably. The numbers
are staggering. Of those drugs that make it as far as clinical
trials, 19 of 20 ultimately fail to gain approval. The success
rate drops to 1 in 100 (99/100 failures) if you go back to the
early preclinical development stages of a potential drug. In
some areas, notably Alzheimer's and dementias, the rate is
essentially zero. The costs accompanying these failure rates
are equally immense, ranging from $200 million to $1 billion
per failure. It is hard to imagine any other activity, commercial or otherwise, staying in business in the face of such high
and costly rates of failure. But they do.

Now I know that the image of Big Pharma is not particularly positive in the minds of many. Of late these big drug companies have been getting a lot of bad press, not to mention piling up liabilities. Some of it is undeniably deserved; they've cost lives. There is no defense for that. But it doesn't have to be this way. It wasn't always this way. In 1990 the drug giant Merck was recognized in a national poll as the most respected corporation in America. Not just the most respected pharmaceutical company: the most respected of all corporations in America. Today big drug companies are about as popular as tobacco and oil companies in the public mind. This isn't altogether deserved because these companies really do work on cures for some very unpleasant diseases and they have certainly produced some very effective medicines. They have also left a legacy for which they rarely get credit. Many over-the-counter drugs you buy today for pennies per pill (packaging and advertising are what you mostly pay for)—Imodium, Zantac, Tylenol, Benadryl, ibuprofen; it's a long list—were originally developed by pharmaceutical companies.

One might hazard a guess that the problem is trying to blend science and business, two activities that might seem to be fundamentally poor bedfellows. But engineering does a fine job of it. Chemistry continues to be profitable and innovative, if perhaps not always well respected. And for some period of time, at least through the early 1990s, apparently biology did quite well. So what happened? No shortage of books and papers have been written on this subject. But my search through the literature turned up a curious bias—it is dominated by market analysts and investors. Very little is written by the scientists working on drug development. There are lots of graphs and tables, but they are all about price per drug and economic trends influencing R&D. I'm a scientist not a businessman, and I certainly wouldn't recommend that anyone take any investment advice from me. But I think there is a perspective that is missing in all these attempts to understand the pharmaceutical business, and I think I can represent it. That's because it's all in the failures.

Pharmacology and drug discovery have been driven by extremely high rates of failure. And by most analyses these rates have not changed in more than 50 years. The number of clinically marketable drugs introduced each year has remained nearly constant since 1950. This is so in spite of dramatic advances in the biological sciences, a greatly expanded disease knowledge base, the advent of genomics, more efficient technologies in chemistry to produce possible drug molecules, techniques that have increased drug candidate screening by more than a hundredfold, and at least 10 times more scientists working on drug development than in 1950. Oh, yes, and many billions more dollars being spent.

The flat lines in the graphs of drug production over six decades might make it seem as if the number of new drugs discovered is a constant of nature, immune, if you will, to outside forces and advances. Attempts to increase productivity or success rates have been resistant not only to advances in science but also to changes in management and organization. The past two decades have seen a rash of mergers and acquisitions leaving fewer than half as many major pharma companies today as there were in 1990. In spite of this, there is no evidence that making bigger companies either increases or decreases the number of drugs produced. This is called the $1 + 1 = 1$ effect. If two companies that each produced two drugs per year merge, then the new, larger company produces about four drugs per year. So size, or more appropriately scale, does not matter. (It is worth noting that there are financial benefits of a merger or acquisition and that's why they keep happening. But the R&D in the companies seems peculiarly unaffected—at least by the simple measure of drug output.)

There is a lot of handwringing about the empty pipelines of the pharmaceutical companies, as if this were something new and that perhaps we have reached some sort of impenetrable biological limit on the targets for new drugs. You know, past performance does not predict future outcomes. But the numbers don't actually show that. They show a steady, albeit small, production of new drugs for decades. In fact all the new technologies are working quite well at keeping the production of new drugs steady in the face of increasingly difficult to solve targets. What's wrong is the unsupported expectation that new technologies and more money should have resulted in increased drug production.

Investors in pharma look with envy at the technology sector and Moore's law, which shows a doubling of microchip power approximately every 2 years. What they see in pharma is the opposite—that the amount of investment money in R&D has doubled every 9 years, but not the number of drugs. To an investor, that effectively means that productivity has gone down—the cost of a new drug continues to increase exponentially, and has been doing so for the last 30 to 40 years. From an investment perspective I suppose this is not good. But from the perspective of finding new treatments it is a remarkable success story. Business and markets are simply not as good at failure as science. There's the rub.

The industry response, because of course it is still an industry, has been to "fix" the imbalance between rising R&D costs and steady discoveries by reducing the R&D budgets. This has the immediate effect of making the drugs-per-dollar ratio look better and therefore makes the investors happier. But limited budgets mean that the R&D divisions are going to take fewer gambles. They will jettison potential but risky drug candidates sooner. They will focus only on big diseases (i.e., lots of patients/customers) that seem tractable. They will, in other words, lose their tolerance for failure. It should be obvious that this is not a sustainable strategy. It can result only in an escalating feedback cycle of spending reductions and fewer drugs.

But I don't think that the individual scientists working in these R&D labs have lost their appreciation for the role that failure plays in discovery. This untenable course is being driven by investors who, just like the federal government that they so often criticize, have no appreciation for the value of failure. They want innovation but don't want to include that cost on the bottom line. Yes, it's true they can take their dollars elsewhere, as they have done in droves. And the executives who run the pharmaceutical companies can cut and reorganize and merge and acquire all they want to appease these investors and promise them bigger returns in the future. But you can't get past the Second Law of Thermodynamics and the entropy invoice that comes in the form of failure.

What's being tragically missed here is that these failures are also positive outcomes. The CEO can, with a good conscience, promise investors that many of their proprietary failures will bring about breakthroughs that we can't yet see, but are surely there. In this case past performance can predict future outcomes. Failure will happen. But these failures are fed back into the company knowledge base, and they will lead to new ideas, and new ideas will lead to new strategies, and new strategies will lead to new drugs, and you will get paid off. Patience.

I said I love Big Pharma because it's the best at failing. And even better, it puts a price tag on it, so we can measure it. Currency is one of the best measures available for hard-to-measure stuff. (I find going to auction houses fascinating because the works of art have a price tag on them. I may or may not agree with the valuation, but there it is, in a number that someone presumably knowledgeable arrived at. Art would seem to be the hardest thing to measure, but auction houses do it by using dollars as the unit of measurement.) To use money as a measure of failure means that most of the cost in drug discovery can be traced to failure. To be sure, there are many who put forward other claims. These include the intense regulatory environment, the long and involved testing required before approval, the requirement that a new drug be better than existing drugs (that's not the case for many other products—you can sell a computer or a car that is worse than or equivalent to others on the market and price it accordingly), the short time a company can profit from a drug before it goes off patent, the lack of so-called low-hanging fruit (i.e., easy targets). I have no doubt that these all contribute to making pharma a difficult business, but none of them has a significant effect on the failure rate. In fact, in some cases they actually serve to increase innovation and discovery. Compared to their counterparts around the world where drug regulations are generally far more lax, American companies still produce more and better new drugs, apparently challenged to do so by the very constraints on them.

So why is it so hard to make a new drug? There are certainly no shortage of pathologies that need treatments. Whether it's kidneys or lungs, heart or liver, nerves or brain, mood or pain, infection or rejection—the opportunities are immense. Targets abound. All the more puzzling that it's so hard to land on one. One hears that the low-hanging fruit is gone—that is, all the easy drugs have been discovered. But that's an old whine in science and it's never been true in any field. One of the small tragedies of science is that grand discoveries become commonplace in a very short time—and then we forget how hard they were to come by originally. The history of every drug available today is full of failures and frustrations and unexpected breakthroughs after a lot of hard work.

I think the problem is much simpler. Biology is still a new subject and it's full of surprises. Evolution is not rational, and if anyone needed proof that there is no intelligent designer they should talk to the chief scientific officers of drug companies. Every failed drug started out as a very good idea about how something worked—whether tumor growth or stroke, depression or diabetes, viral infection or emphysema, or, or, or . . . But things aren't engineered the way we would expect them to be if they were designed rationally, and thus biology fools us at every other step. One is simultaneously humbled

and proud. Humbled in the face of our unbounded ignorance of biology and proud in our ability to decipher even
the small bit of it that we have. It is critical to be cognizant of
our ignorance, and our failure rates are simply a very good
indicator of the bounds of that ignorance.

This is no cause for despair. Failure never is, as I hope that
theme has become clear by now. There are many new opportunities in the drug development field precisely because of
all the failures. For many years immunotherapy was ignored
because it had failed too often and was thought to be simply too hard to figure out. But then the vast microbiome of
organisms living inside all of us was uncovered, and this led
to ideas about novel and previously unconsidered functions
of the immune system. Now many of those previous failures
made sense and suggested new ways to deliver immunologi-
cals as drugs. And many of these immune-based therapies
are being tried with surprising, if patchy, success. They still
fail a lot, but every failure gets us closer to understanding
how to make them more dependable.

Contrary to the common thinking that ever larger (and
more expensive) randomized trials produce more statistical
power, it now appears that using smaller patient bases often
reveals more details about why something fails. There is more
to failure than so-and-so drug failed to produce more than a
35% improvement in a large clinical trial. Large drug trials
produce reams of data but in the end a more or less binary

result at some predetermined level of significance—yes it works or no it doesn't. But there is much more to ask. Who did it work for and why? Who specifically didn't it work for and why? Did it work partially in some cases or for a while and then stop working? There are all these new and unexpected ways of failing, and every one of them is a step closer to a new wonder drug.

Failure is the frontier of pharma.

FOURTEEN

A Plurality of Failures

But I may think otherwise tomorrow.
—Joseph Priestly, announcing his discovery of oxygen

On November 7, 1997, the *New York Times* carried a full-page obituary, running to some 4500 words, on the death of philosopher Isaiah Berlin at age 88. The *Guardian* of London had an even longer piece on Sir Isaiah, the knighted philosopher from Oxford. Berlin had championed a moral, ethical, historical, and ultimately political system of thinking. An approach that he called pluralism, or, more technically, *value pluralism*. He carefully distinguished his value pluralism from the relativism and subjectivism that enjoyed some popularity among philosophers and other humanists of the time.

Berlin's value pluralism was at once far more radical and more constrained than either relativism or subjectivism. It was not "anything goes" but "many things go"—or better, "many chosen things go." Berlin claimed that there were values that were both good and incomparable, or, to use his better word, incommensurable. That is, two or more things could be valuable or good, and yet not be measurable against each other, nor could one decide between them on a purely rational basis. Worse, they may even be in conflict. Liberty and privacy might be an example we struggle with today. Both are recognized values, but they cannot be measured on a single scale and compared with each other in the same units (what you would pay for them, for example), and they are often in opposition to each other. Nonetheless, choices must be made. Unlike relativism or subjectivism, the value of one thing or another is not a matter of opinion or personal perspectives, or even context. The differences are real and objective, and incommensurability is an attribute of both. We must therefore make decisions in the face of incomparable and sometimes antagonist values and goods.

This may all sound existentially dreary, but Berlin saw this state of affairs as exemplifying a human condition of glorious pluralism. The feat was to recognize the value of pluralism, to engage it, to expand it, to let it flourish in a truly liberal society. "There is more than one way to skin a cat," and no method is necessarily better than the others. (I don't know why it's a cat. Personally, I like cats and would not skin one.)

Pluralism is a creative force because it admits of multiple ways to see a thing, multiple valuable paths to choose from.

Having different choices, especially difficult and incommensurable choices rather than just variations on a theme, is expansive and stimulating. Even after you make some choice, there remains the likelihood that others will make different choices and so come to findings in a different, and possibly enlightening, way. We need not all take the same path, and doing so would be socially totalitarian and personally uncreative. Berlin essentially denies that there is a single correct way to do or see any human activity and asserts that sometimes the multiplicity will create rational or logical inconsistencies. Live with it. Live well with it.

One of Berlin's most famous books was really an essay called the "The Hedgehog and the Fox." It was written as a critique on Tolstoy, and in it he used a line from an ancient Greek poet, Archilochus—"The fox knows many things, but the hedgehog knows one big thing"—as a classification scheme for writers, thinkers, artists, and others. Although it is arguably the most popular of his writings, and the most often cited, Berlin himself said, "I never meant it very seriously. I meant it as a kind of enjoyable intellectual game, but it was taken seriously. Every classification throws light on something." And it is a singularly useful shorthand for differentiating between monistic and pluralistic attitudes. In the essay he gives examples of each kind of thinking, including

Plato, Dante, Pascal, and Nietzsche as hedgehogs and Aristotle, Shakespeare, Montaigne, and Joyce as foxes. Of course there's usually a bit of both in most of us—even Tolstoy, he concludes, was by nature a fox but by conviction a hedgehog. Pluralism, though, is the province of the fox.

Berlin applied his philosophy of value pluralism to art, literature, history, politics and ethics, but largely ignored science. I think that was perhaps simply because it didn't interest him enough and he lacked, or felt that he lacked, sufficient expertise to do so. It was just not his battle. But that is no reason to believe that science enjoys a simple monistic set of beliefs that are indisputable and immutable. Far from it: science is filled with ongoing mysteries and unexpected findings, and, most important, apparent paradoxes from which the most creative solutions emerge. In other words, interesting failures. There is, then, no reason that Berlin's value pluralism cannot be extended to scientific activities. Let's go to the lab and see how pluralism can be of value there, too.

Science is a method for observing and describing the universe. It can be monistic only if you first believe that the universe is essentially monistic. If you are convinced that there is finally a single overarching explanation that will describe the universe in its totality, now and forever, that this principle or mathematical formulation is discoverable and will be discovered, then you may, at least logically, take the monistic approach. Although, practically speaking, it still seems to be a very limiting perspective. Even if ultimately there is a single encompassing Truth, we don't seem to be anywhere near ultimately in any of our sciences, so why behave that way? And it should be noted that although many scientists and vast swaths of the public likely believe that the "real" world is a single entity explicable by a few fundamental laws, there is not one shred of scientific evidence that this is so.

On the contrary, there are highly notable cases where a monistic view is simply incompatible with the evidence. In physics the battle to decide if light acts as a particle or a wave has been abandoned in favor of the dualist view that it can be either or both. And Heisenberg insists that a fundamental feature of matter is that the elementary particles of which it is composed require a certain indefiniteness in their measurement. Even in mathematics, carefully constructed from axioms and logically derived inferences, there are Godel's incompleteness theorems showing that a single answer cannot be proven to be the only answer within any logical system. Biology may not have such conundrums currently, but we have yet to resolve some of its more difficult problems—consciousness, development, even evolution. Any one of those could hold incommensurable facts. Altruism springs to mind as an example of what remains an evolutionary conundrum.

Even if there were a simple explanation to be had, it's not clear that it would be easy to understand or comprehend.

Einstein's $E = mc^2$ seems like a pretty simple mathematical formula, and yet its vast implications are hardly understood by more than a handful of people. I was at the Nobel Prize lecture of physicist Frank Wilcek, who pointed out that even the simple algebraic manipulation of that iconic formula to $m = E/c^2$ reveals a view of mass that wouldn't be obvious to most very smart people (that a body's inertia is a function of its energy content). Simplicity, even when it's available, is not a substitute for pluralistic thinking.

Where does science fall in Berlin's hedgehog-and-fox classification scheme? First, let me distinguish two kinds of science that are going to be relevant to the discussion here. There is the personal, at-the-bench science that is practiced on an everyday level—doing experiments, discussing results, identifying and solving problems, writing papers. Then there is the wider culture of science to which every scientist belongs and is engaged with to varying degrees—teaching, reviewing papers and grants, serving on committees. Hedgehogs and foxes can be applied to both, and sometimes, like hedgehog and fox behavior, they morph into each other.

The personal science is well captured in a curious linguistic construction scientists use when they talk about their individual work. They often refer to it as "my science." "My science uses genetics to explore breast cancer." "My science analyzes patterns of electrical activity in the nervous system." Lawyers don't say, "My law is about the

First Amendment," or "My law is about corporate regulation." They don't have a "my law." Even doctors don't say, "My medicine is cardiology or neurology" or whatever their specialty is. They don't have a "my medicine." Scientists have a curiously personal and proprietary relationship to their work. While charming in its near childlike enthusiasm, I believe this tends to breed a monistic, hedgehog perspective.

Some part of this monistic approach is due to science's heavy reliance on technology. A laboratory becomes proficient in certain techniques and procedures or using certain tools. A particle accelerator or an electron microscope, molecular biology or anatomical staining, stem cells or satellite telemetry, and so forth. Every discipline, and every subdiscipline, has its set of sophisticated tools or analytical techniques. Mastery of these tools comes at a fairly high price and is a considerable part of every scientist's expertise. It seems that the scientist knows one big thing and forages around for tasty questions like a hedgehog.

Being careful to heed Berlin's warning and not take this metaphor too narrowly, scientists don't just lumber around and shine their X-scope on one thing after another just because they can—at least most don't. It still takes imagination to think of the experiment—the apparatus, complicated and sophisticated as it may be, does not finally think of the experiment. And consistent if not persistent failure, because you are using the newest, most sophisticated, and therefore the most unreliable, technology, forces you to be more foxy.

Aside from the inherent monistic perspective that your toolbox imposes, there is also the dedication to an idea, the focus on a question, the consuming passion to find out about something in penetrating detail. All of these reinforce a monistic, hedgehog outlook. I don't want to call it myopic, because within a certain specified area it may be very wide ranging, but there is a tendency to believe that all the relevant questions can be approached and answered by the judicious use of your expertise—your one big thing. But then inevitably failure intervenes and forces you to be more foxy. When the tried-and-true technique won't give up the solution, you must consider alternatives, even those that may fly in the face of good sense, your prized hypothesis, or supposedly settled facts. So the hedgehog becomes the fox, for a bit.

Let's now contrast this with the role of the scientist in the culture of science. You may be surprised to learn that senior scientists spend fully half of their time on issues outside the lab but that are crucial to the infrastructure of science. They are called upon to sit on government or university committees that deal with everything from academic policy to room assignments; they review grants and papers; they are responsible for hiring, promotion, and tenure decisions; they determine the curricular requirements for undergraduate and graduate students; and they oversee graduate student admissions. In all these and other ways, a critical part of their job is to support the institution of science. And it is in these areas, if not in the direct running of their own laboratories, that monism is especially damaging and where pluralism will be of paramount importance. If some senior scientists believe that molecular genetics is real science and behavioral psychology is not, then their decisions about all of these infrastructure issues will be monistically reflected in their choices and votes.

As an example, many scientists consider it their duty to decline requests to review papers or grants that are too far outside their area of expertise. I don't agree with that; it's too monistic a view. Reviewing and judging requires its own kind of competence, and it requires as much expertise as doing experiments. A good scientist can tell a decently written grant from one that is simply full of crap. A good scientist can tell a legitimate question from an idiotic one. Even if it's not in his or her immediate field. Not with 100% accuracy. But half the Nobel laureates can pull out initial rejection letters from so-called expert panels at granting agencies and journals for work that later earned them the big prize. So, even with experts, it's not perfect, and the errors, errors of omission, may be more damaging to progress than the bit of waste that might occur from funding a few bad science projects through misjudgment born of open-mindedness and fresh perspectives.

Similarly, deciding on where to put research resources would be best served by a radically pluralistic process that maintains as many options as possible while a problem is still unsettled. It would be very sad suddenly to find yourself face to face with an insurmountable problem and have no alternatives that are developed enough to progress in new directions. One of the most important values of pluralism is to make failure, even big failures, noncatastrophic. If one thing fails, it will not cause a widespread collapse. Charles Sanders Peirce, the 19th-century American pragmatist philosopher and scientist, uses the wonderful metaphor of science not as a chain of findings and methods that is no stronger than its weakest link, but as a cable composed of many thin threads, numerous and intricately connected. The failure of a few of those threads will not substantially weaken the strength of the cable.

Adopting this kind of value pluralism means that we must accept a significant risk of failure. Just as all the different investigative means could lead to success, so it is possible that only one will, and the others will have been a waste of time and money. Can we avoid this? Only by putting all our money on one horse, which is not a good strategy, as any successful investor will tell you. The cost of pluralism is tolerance and patience for failure. What we get back in return is a richer, more inclusive, more intellectually engaging science. Not a bad trade, in my opinion. Yes, it will almost surely cost a bit more. Still a good bargain.

The field of neuroscience is a perhaps typical, although not unique, example in modern science. It suffers from a plurality of competing monisms, which is different from true plurality. Among neuroscientists there are those who believe that studying the brain in any animal "lower" than a monkey is a futile exercise. Others believe that bothering with any creatures except mice, an animal whose biology happens to make it possible to perform sophisticated genetic manipulations, is just wasting resources. Cellular neurobiologists believe that you can never understand the mind, if such a thing even exists, while cognitive neuroscientists believe that you can know everything, everything, about single neurons and it will tell you almost nothing about how the brain works. And they are all valuable views! That's the thing about pluralism: they are all correct, and they are all strategies that should be supported. And then we will have a vibrant brain science. Does anyone really think it is likely that something as complex as the nervous system will be explained by a single principle?

Rodney Brooks, the Australian roboticist best known now for his friendly robot Baxter, has been rethinking what robots could be for decades. Not only is he a pluralist, but his robots are pluralists. In one case he recounts, NASA had a contract offer for building a robot to send on one of its missions. But the weight constraints were difficult and required cutting corners so that ensuring that the robot would actually work was risky. Instead, Brooks proposed making hundreds of tiny robots, with the same aggregate weight, that could accomplish a variety of tasks and sending them out like so many insects once the probe landed. Even if many of them failed, there would be a sufficient number to gather useful data.

And speaking of insects, Mark Moffett, Smithsonian Fellow and entomologist (his website, known as Dr. Bugs, is immensely informative and entertaining) points out that ant colonies work on this pluralistic scheme. While they may exist as a super-organism with billions of members carrying out their individually very monistic duties, when they need information about the outside world—food or foe near the colony—they send out millions of individual ants, most of whom either will never make it back to the colony or find the needed information. But enough will so that the colony lives on successfully. Very successfully. Transforming yourself from hedgehog to fox and back is a very neat trick. We could learn some things from these ants.

From an even more improbable source, the annals of monotheism, there is a Talmudic story that captures the deeply almost paradoxical nature of pluralistic thinking. The Rabbi, teaching a portion of the scriptures, asks the first student to interpret the passage. After the student presents his involved interpretation the Rabbi says, "Very good, you're right!" He then calls on the second student, who proceeds to give a diametrically opposed but equally detailed interpretation. "Very good," says the Rabbi, "You're right!" A third student protests, "But Rabbi, they can't both be right." "Very good. You're right!" says the Rabbi.

Pluralism in science is an idea of growing interest and popularity in the philosophical literature. The philosophy of science does not always have direct relevance to the everyday practice of science, and indeed it is often ignored, if not outright disparaged, by scientists. This is an error on their part, in my opinion, but that is a topic for another time and place. However, pluralism and the new ideas associated with it could and should have a very direct effect on both the practice of "my science" and in particular on how we operate this activity called science.

The pluralistic approach provides a sturdy foundation for the uncertainty and doubt that are often endemic to science. Pluralism accepts the value of ignorance and failure in the process of gaining scientific knowledge. It is how uncertainty, doubt, ignorance and failure are transformed into a force of creativity rather than a source of discouragement. It makes room for a multiplicity of theories and approaches in science, even ones that are not entirely correct. Pluralism recognizes that some things will be known with near certainty while others can remain at least temporarily uncertain. It tells us that unsettled science is not the same thing as unsound science and requires us—not just scientists but all of us—to work hard on making that distinction. Precisely because we admit that alternative explanations can coexist—at least for a time—it becomes ever more important to understand that a viable explanation does not have to be *the* explanation.

In a pluralistic scheme one still makes choices—thus *value pluralism*. Not everything is equal. Importantly, Berlin recognizes that equivalence of ideas is not a part of pluralism. That would be the simpler and less rigorous condition of relativism—the currently fashionable "whatever." But in monism one makes choices as well—more extreme choices because everything but one way is dismissed. Pluralism means many, but not any.

And once the gate is opened, I hear you protesting, how is one to decide what to let in and what to keep out? But that is a shallow argument born of a kind of intellectual laziness that would prefer one single way that is correct and identifiable, thus removing the messiness of the world. Fine, but removing the messiness also removes the richness. Value pluralism means that one still makes judgments. But those judgments don't have to come down to a single winner. Even if all roads lead ultimately to Rome, multiple paths offer a more interesting range of journeys than everyone clogging up one motorway. And an accident on one route does not lead to a standstill in Roman commerce. Anyone who has driven in Rome may understandably find this analogy far-fetched, but you get the idea.

How does science stack up against other human endeavors, other epistemologies, from this perspective of failure and pluralism? Business, economy, government, the arts? The arts, it seems to me, are most parallel with science. They require failure and ignorance and uncertainty and doubt to fuel creativity and discovery—they are best when they maintain pluralism, although it is often the case that they are taken over temporarily by fad or fashion. Business seems fundamentally different. In business the winner-take-all ethic is predominant. Success is achieved by devouring the competition, not tolerating them. Monopoly (there's that *mono* root again) is the ultimate victory in business, thus the need for legislative intervention to suppress them. I know there are endless volumes of business books that extol failure and risk taking, but it's all in the service of a strategy for world dominance.

Although Isaiah Berlin makes a compelling and impassioned case for pluralism in government and culture, I personally see little of it in modern discourse. Politics has certainly become a winner-take-all game. The original ideas of pluralism associated with liberal democracies and legislative deliberations have all but vanished in favor of the fervor of one true path. Voters have become believers, not choosers, and alternatives are dispatched with smug righteousness. There are still sides, and one can choose to be this or that (I would have said conservative or liberal, Democrat or Republican, but these terms have lost their meaning and are labels only for belonging to a particular party line that believes completely in its supremacy). These days joining a side merely means spouting their slogans. There is no good failure, no failure with value, in politics or governing. Thus the mess.

Technology, which straddles science and business by way of industry, is an interesting example of a kind of chimera. While it may initially be pluralistic, in its scientific and developmental stage, it eventually becomes monistic. When videocassette recorders first appeared on the market in the early 1970s, there were different formats to choose from. Each had valuable features but they were not compatible. The consumer had to choose between Sony's Betamax and JVC's Video Home System (VHS). A format war of sorts developed (this is now a classic case study in business schools), with VHS eventually winning and driving Betamax out of existence within a decade. There were many good ideas embodied in both systems and many other good ideas likely would have come from continued research in both formats. But they were, in a very literal way, incompatible. The multibillion-dollar industry that controlled the technologies was not interested in pluralistic values but rather in very absolute currency. And Betamax, with all its technological advances, disappeared without a trace.

An even more telling and longer lasting battle has been raging in the personal computer industry. Far too much has already been written about the fight to the death between the Microsoft and Apple operating systems. I wish to point out only that having more than one operating system for all the personal computers in the world is not likely to be a bad thing. But apparently as far as Apple or Microsoft is concerned this is tantamount to apostasy. Because of this monopolistic thinking we were within a hair's breadth of having only one operating system survive—Apple was very close to disappearing. Perhaps in reaction to that, there are new players in the field (Google, Linux) and the possibility that operating systems will remain pluralistic. Really, though, could it have ever been the case that there was only one good way to run a computer? And even if there were, what is the likelihood that the first one (or two) we came up with would be the best? Yet this is the argument that Bill Gates or Steve Jobs would have made. And perhaps their heirs still think it's worth making, along with too many other Silicon Valley winner-take-all capitalists masquerading as captains of innovation.

We could compare religion here as well. But I'm guessing that no one will object to my avoiding that discussion. When the British Royal Society was founded in 1660 as the first of its kind Academy of Sciences, among its charter rules was that there were to be no discussions of religion or politics. No doubt this rule has contributed to the longevity of the Society. I have already made the mistake of including a paragraph about politics. I believe I should avoid the almost surely fatal error of engaging religion.

Not hearing any objections (because I'm writing this book alone in my apartment), I'll take it that you all now believe pluralism is desirable. But we should be clear that it isn't a requirement. If something is known and established to the agreement of all concerned experts, then it is probably counterproductive to develop new models simply on the basis that you can't prove it couldn't be some other way as well. Admittedly, I can't prove that the positions of the planets in the night sky don't influence my love life—but because there is no evidence to support it and lots to disparage it, there is little reason to pursue the issue.

At the same time, having something apparently settled to everyone's satisfaction doesn't mean that everything else is now useless and can be discarded. After all, many important things that once seemed established—absolute space and time, biological vitalism, even atoms—have been shown to be incomplete or even incorrect explanations that needed to be changed, if not replaced, years or decades after their general acceptance. But the bar is raised in cases of strong models or theories, and the admission of new systems requires strong evidence. This is not monistic, just sensible. It is in fact why pluralism can work, as distinct from relativism and subjectivism. Pluralism is more difficult than monism, relativism, or subjectivism. It requires us to be more tolerant and more suspicious at the same time. Ironically, it recognizes the incommensurable values of tolerance and wariness, of broad-mindedness and skepticism.

. . .

This has been a long chapter, so let me summarize and try to make a clear connection between pluralism and failure. Failure, ignorance, uncertainty, and doubt are all critical ingredients of science. If you accept that diversifying and hedging bets are good ideas in the face of ignorance, uncertainty, doubt, and potentially high rates of failure, then pluralism is for you. Looking for a lost camper, who clearly has only one, if unknown, location, is nonetheless not likely to be successful by sending everyone off down the same path. Fanning out is the method of choice.

But pluralism also makes failure more likely. Fanning out may be the best method, but that means that most of the searchers won't find the camper. Similarly, in most scientific searches there may very well be a single or small number of correct answers (see chapter 5, "The Integrity of Failure")—and they may turn out *not* to be entirely incommensurable. Will all of science be reducible to physics, as the somewhat hubristic physicists would have you believe? I don't think that's the case, although I can't tell you for sure. But even if it is, we won't find that out by studying only physics.

I have described two perspectives that science may adopt in its ongoing search for explanation—monism and pluralism. Clearly I am in favor of pluralism. Further, I am claiming that a fundamental principle for practicing pluralism is an acceptance of failure. I might go further and try to claim that failure is a requirement for pluralism, and even that acceptable, noncatastrophic failure enables and perhaps even facilitates pluralism. Or is it the other way around? Or is it both ways?

I have been a follower of Isaiah Berlin's writings for some time, and have always had an interest in seeing them applied to science. I was fortunate enough to find a doorway into that quest as a guest of the Department of the History and Philosophy of Science at Cambridge University, and in particular through my friendship and many discussions with Professor Hasok Chang and others he introduced me to. My comments about pluralism follow those of several philosophers and commentators on science, including Nancy Cartwright, Stephen Kellert, Helen Longino, Israel Schleffer, Elliot Sober, and numerous others. I have avoided the arguments for pluralism by such philosophers as Thomas Kuhn and Paul Feyerabend because I think they are more complicated than would be appropriate for this book (and this author); also because I personally think their ideas are not commensurable, if you will, with those presented here, and are frankly less relevant. I have not included direct citations as this is not an academic publication and I preferred not to interrupt the reader's flow with intrusive notes and citations.

Instead I have included extensive reading lists in the notes for those interested in pursuing this fascinating avenue further.

EPILOGUE (IN CASE YOU HAVEN'T HAD ENOUGH): A CASE STUDY IN SCIENTIFIC MONISM

Charles Darwin believed quite strongly that animals had mental lives, including thoughts and emotions that were different in quantity but not quality from those of humans. He owned a big Newfoundland dog named Bob. I too am servant to a gently insistent Newfoundland (named Orsin) and so understand why Darwin thought animals had a mental life. But besides matching wits with a Newfoundland, Darwin believed, because evolution would suggest it, that there is no reason to think that the mental aspects of life obeyed different rules than any other physiological or anatomical or biochemical aspect of life. That is, there is an evolutionary continuity between all living forms and functions as expressed in the myriad extinct and extant species. If one can track the continuous evolution of the cardiovascular system, then there is no reason to suspect that the human nervous system was unique in having sprung up spontaneously. Although Darwin may have had little empirical data for this, he believed it as a matter of an almost commonsense proposition following from the principles of evolution.

Following Darwin, a field of animal behavior known as ethology emerged, especially in Europe. Ethology was the investigation of animal behavior through observation and experimentation, primarily in the wild but also among captive or domestic animals, with the goal of understanding their behavior in more or less natural contexts and through the lens of evolution. The great leaders of this field were Konrad Lorenz and Niko Tinbergen, joint winners of the Nobel Prize in Physiology or Medicine in 1973.

Ethology was based on the naturalistic study of animal behavior and unabashedly employed the belief that animals possessed mental states that we could understand empathetically. This is known as *anthropomorphizing*, a bit of a mouthful, and now something of a dirty word in behavioral science. It means literally interpreting the external behaviors of animals as you would those of a person. That is, as emanating from mental states—desires, fears, drives—that are not directly observable but can be intuited from our own mental experiences. While this is done with human subjects in psychology experiments all the time, there is the advantage that you can ask humans for a verbal report of their mental state while you cannot do this with animals. How reliable that verbal report may be is anyone's guess and has become a rather controversial issue of its own—in psychology, economics, and jurisprudence. Nonetheless, you can't ask an animal to report its mental state, not directly.

Ethology flourished, especially in Europe, and produced some new ideas and new ways of thinking about old problems—imprinting, instinct, animal learning, social behavior, nature versus nurture, and so on. It was by temperament a low-tech kind of science that fit quite well with other branches of biology around the early 20th century. In the United States, scientific interest in behavior took something of different path, mostly through the writings and influence of John B. Watson, one of the giants in American psychology. He introduced the idea of behaviorism, which stated that the only true subject for the scientific study of behavior was through what could be observed, not inferred.

And then in the 1960s a young psychologist, soon to become famous or infamous, named Burrhus Frederick Skinner, or BF as he understandably preferred to be called, took up the behaviorist flag and instituted a rigorous experimental program. In a direct rejection of the ethology school of behavior, he proposed a radical dismissal of what could not be directly observed—which meant everything except overt behavior. Mental states were at best conjectures and could not be the subject of true scientific investigation. Further, radical behaviorism, as it came to be known, claimed a scientific high ground by being entirely experimental, with no attempt to even approximate a natural context or environment—physical or social. Skinner is perhaps most famous for inventing a cage-type enclosure, later known as a

Skinner box, for housing pigeons or rats or mice. (Most infamously, he also built one for his daughter to play around in.) In this controlled environment, food rewards were doled out based on the behavior of the animal in the box, or on the particular contingencies of the experiment. The point was that the environment was under the nearly complete control of the experimenter and every behavior that the animal performed could be observed and carefully noted, thus producing huge amounts of data. It had all the trappings of real science.

Let me use two classic experiments to contrast the approach of each group. In the early 1900s, ethologist Wolfgang Kohler investigated the mental abilities of chimpanzees by observing them in the wild and finally running a famous experiment on zoo-housed chimps. He placed a bunch of bananas high out of their reach and left some wooden crates lying around the enclosure. After some time the chimps piled the boxes on top of each other, climbed them, and reached the previously inaccessible bananas. Kohler claimed they came to this solution with little or no prior experience but rather as a matter of novel insight—a sort of "Aha!" moment of discovery by mentation.

BF Skinner, in 1947, using pigeons in one of his experimental boxes, performed a clever experiment in which he claimed to have developed superstitious behavior. A food pellet was delivered to the pigeon every 15 seconds, no matter what the animal was doing. That is, its actual behavior was completely irrelevant to obtaining the food "reward." The bird was left under these conditions for "a few minutes every day" and after some days (the number is curiously not specified in the 1948 publication of the experiment) each of the several pigeons, in their own cages, could be seen performing a very distinct set of behaviors leading up to the delivery of the food pellet. One hopped around from one foot to the other, another thrust its neck in and out, another turned circles—each one developed a set of behaviors that they performed until just before the feeding mechanism was activated, and then they would go and get their reward. Skinner observed that these behaviors were the result of the bird performing some random behavior when the pellet was first released and that this particular behavior was therefore reinforced. Repetition of the behavior continued to result, from the bird's perspective, in the delivery of food so the behavior was further reinforced. Skinner claims that these behaviors, unconnected to the reward in fact, are very much like superstitious behaviors people develop in situations when their actions cannot actually alter the course of events. Gamblers or athletes with various "lucky" rituals come most easily to mind.

I find both sets of these experiments and observations fascinating. They both shed some light on behavior and its source. They both provide tantalizing if incomplete models for complex behaviors that are not uncommonly found in humans as well as other animals. They both illuminate among the most difficult of subjects to study—the brain and behavior. And yet adherents to these two schools—ethology and radical behaviorism—will not even talk to each other. They cannot even be civil toward each other. There are vicious attacks on each other's work in the press and scientific literature, reports of promotion and tenure denials by departments dominated by one or the other of the two schools. Indeed, virtually all psychology departments were at one time, if not still, known as belonging to one or the other schools. Radical behaviorism has even come to be known more commonly as Skinnerism and its adherents as Skinnerians, as if it were a cult or nationality. A most remarkable state of affairs for people studying behavior and its consequences. Unfortunately, the irony seems lost on the participants.

The behaviorists accuse the ethologists of anthropomorphizing as if it were a psychological disease. The ethologists claim that operant behavior is just a set of circus tricks souped up to look like real behaviors but which are completely restricted to impoverished laboratory-type settings. Neither believes the other is practicing real science. Neither believes the other has anything valuable to offer neuroscience. For quite a long time the radical behaviorists, led by Skinner and his many students, dominated American psychology. Funding for ethological work was nearly impossible to obtain from NIH or NSF. Characterizing someone's work as anthropomorphic was fatal and immediately marginalized the person and the work. Almost the opposite landscape prevailed in Europe where ethology was considered to be intellectually superior to the "rat runners"—a snarky reference to using laboratory-bred rats for behavioral experiments in mazes and Skinner box–type arrangements (although Skinner himself always preferred pigeons).

I can't think of a worse case of monism in the annals of modern science. Here are intelligent researchers trying to understand the most difficult thing on Earth to comprehend—the brain, its perceptions and behavior, how it (we) thinks—and they can see it only as a chauvinistic fight to the death. They train cadres of students as if they were armies, they publish papers in their own journals as if they were propaganda instruments, they have separate conferences as if they were political conventions. We have a full-blown schism here—in many ways uncomfortably similar to the divide between Catholic and Protestant Christians or Sunni and Shiite Muslims. Is there any sense to this? Is this how science should be practiced?

I personally find the experiments and results of both groups enlightening. Skinner and the behaviorists have shown that the brain is easily conditioned to rewards, perhaps dangerously easily, so that we, society, can take control of the situation and set up useful incentives—or let it be random and accept the consequences. They have demonstrated convincingly that the brain is not very successfully conditioned by punishment and that this is an inefficient way to alter behavior—not that anyone in power appears to listen to this useful data. The ethologists have shown us that mental states can be tested and that they are worthy of our investigation. They have shown us that many animals may possess highly cognitive lives and that they deserve our respect and ethical treatment. They have successfully investigated social behavior and the possible origins of puzzling behaviors such as altruism, cooperation, empathy, and friendship. They have shown us that evolution shapes behavior, but does not necessarily determine it. They have shown us that humans are not at the top of the ladder because there is no ladder, just different ways brains solve problems. And these examples are just a small taste of what is contained in these two separate literatures.

Even more interesting are the questions that arise from these research efforts. How plastic—that is, adaptable—is the brain? How much is hardwired, and how much can be molded by learning and experience? What is the nature of communication? How does it differ (or not) from language? Is there a sense of self, and what does it mean to possess (or not to possess) one? Are there limits on learning? How much of learning is conscious, and how much unconscious? What is the difference between learning *what* and learning *how*? And thousands more.

This abominable state of affairs is a very Western condition. The irresolvable conflict between the European and American schools is nowhere to be found in Japan, for example. The study of monkeys and apes has been a leading area of science in Japan for more than a century, producing several internationally renowned primatologists over that time period. Among them were Shunzo Kawamura, and Masao Kawai initially working under the guidance of Kinji Imanishi. Kawai in particular has had a significant influence on the field of primatology in Japan for several decades. He and his associates showed that a learned behavior in a troop of wild monkeys could be transmitted generationally. This is the now famous story of potato washing, a behavior that was started spontaneously (Aha?) by a young female in the troop and spread rapidly among the younger members and then to the next generation by observation and trial and error (conditioning?); the older monkeys interestingly never picked it up (bah, humbug?).

This discovery and others were made—or as in this case, noticed—because of the particular mindset of Japanese primatologists like Kawai. They naturally empathized with their primate subjects to gain a clearer understanding of their behavior. "To understand the monkey, you must understand the monkey's mind," was the credo. But to do this you "record just what you see." This strategy, a kind of intellectual combination of ethology and behaviorism, was unique to Japan.

Why? One theory is that, because Japan was not a Christian (monotheistic) culture, there was never any notion that man was discontinuous from the other animals. Darwin's theory of evolution was widely and immediately accepted in Japan as being almost self-evident—of course there is continuity between all living things and it is not inappropriate to believe that monkeys may share some of our mental experiences.

This attitude, or rather the lack of the negative one, led to the regular use of anthropomorphism in observing and describing animal behavior. In that scientific culture it was not objectionable to apply human qualities to the behavior of other animals. We might see this as a kind of behavioral teleology—that is, the use of purpose to describe function. Teleological explanations are generally regarded as scientifically bankrupt—stones don't move because of their desire to get to another place but because of impersonal forces applied to them. Teleology can lead to just-so stories—giraffes getting longer necks because they want to reach higher branches and the like. But teleology and anthropomorphism, although fundamentally wrong, have some value in directing our minds to things we might otherwise miss entirely. The discovery of large-scale social groups among certain primates or in dolphins, and the "purpose" of forming and breaking friendships and alliances, is best described, at least for now, in anthropomorphic and teleological terms. We have confidence that eventually we will be able to understand them on a genetic and evolutionary basis as well, or as the result of hormonal levels produced in reaction to stress or sexual cues, or some physiochemical mechanism. But until and if we have an ultimate reductionist cause for a sophisticated behavioral activity, why not use the shorthand of teleology? One can appreciate that ultimately teleology and anthropomorphism will not provide a satisfactory answer, while at the same time using them as tools for making new and useful, and true, discoveries.

My point in this long, but I hope engaging and enlightening cross-cultural story, was to show the value of pluralism and the destructiveness of monism in science. All of these methods had much to offer—and each of them failed in many ways. Failures are no reason to jettison a set of scientific data and interpretations. They are very useful constructs, and they help us understand hard-to-understand stuff—even if only provisionally.

FIFTEEN

Coda

Failure is a favor to the future.
—Rita Dove, U.S. poet laureate (1993–1995)

When I started writing this book I had a few clear ideas about failure and its value in the pursuit of scientific explanations. What surprised me was how quickly those few ideas generated dozens—no, hundreds—of questions. I don't know why I was surprised, though; that's exactly what happens in science—answers beget questions, always more questions than answers. And so there are still lots of questions. But you have to end somewhere. It's the only way to get on with it.

I have many regrets about this book. In life regrets are not good, but in books I think they are. There are another 20 chapters, essays, that I could have written or did write but didn't include here. And if I had, there would be 20 more;

I'm sure of it. That's good. There's more to say, more to think about. And of course there's nothing to stop you from going on and thinking up your own chapters. I hope that's what happens. You may have noticed that I have a fondness for quotes. It's not so much that I think quoting famous people gives a statement more authority or is less open to criticism. Rather, I think that just because you're dead you shouldn't be left out of the conversation. And more importantly, we should realize the conversation has been going on for some time. It didn't start with us, and it won't end here. I certainly wouldn't want this book to end the conversation. No conclusions, please.

All that notwithstanding, there is tucked in here a kind of proposal. A proposal for how to think about and even how to run science in this complicated place that science has helped to make complicated. My proposal is that the scientific method—the real scientific of method of welcoming doubt, uncertainty, ignorance, and failure into the enterprise, the method where science is a process and not a pile of facts, a verb not a noun—that this scientific method is not owned by an elite cadre of PhDs and experts. I believe in experts and I am indebted to people who have devoted so much of their time and life to becoming knowledgeable in narrow but indispensable areas. We need dedicated experts to make progress. What is dangerous is when those experts, who know the importance of failure and of doubting and who regard uncertainty and ignorance as opportunities, either hide those facets of the process or simply fail to make them explicit. That is when they become, purposely or by accident, elites. That is when the culture at large feels shut out of science. That is when science engenders suspicion and resentment. That is a terrible failure, the wrong kind of failure.

But if expert scientists are going to talk honestly about failure and uncertainty and doubt, then there is a corollary responsibility for the public who enjoy the fruits of science to understand how these represent progress and not a reason to be suspicious of scientific findings and opinions. Expressing doubt and uncertainty should make a person more trustworthy. Anyone who claims to know the truth, the Truth, because they are special or have a special connection to some authority no one can question, these are the people you want to be wary of. Poet André Gide advised, "Seek out those who search for the truth. Flee from those who claim to have found it." Science is the best method I know for being wary without being paranoid.

So it is also up to that public to understand how science works and what can reasonably be expected to come from it. It is up to that public to do the hard work of being citizens in a democracy and not simply deferring to authority. It is up to citizens to learn an appreciation for failure and ignorance the way a professional scientist does. Because that part of science is available to them without obtaining a PhD.

I'm not saying that it is easily available. I grant that it can be hard to make critical decisions when there are many options, often conflicting and backed only by unsettled (but not unsound) data. It is hard to accept that our success depends on failures and that we need to have patience. It is hard to recognize that it is not output that we should measure, but outcomes. It is hard to fail again and again with no lack of enthusiasm. Mostly it is really hard, as Richard Feynman says, to not fool ourselves. But all these things can be done smartly and thoughtfully with the right attitude toward ignorance and failure. I know this because it is easy to see the disastrous results—in politics, in society, in education, in science—when people claim to know something for sure, when they claim infallibility, when they claim authority. Likewise it is easy to tally the unreasonably successful results that again and again follow doubt and questioning and failure.

Science is a great treasure and an immensely engaging adventure. It works best in democracies and worst in the service of empires. That alone should tell us something about why we value it. It is generational, arriving on our doorstep from preceding generations and being sent off by us to the next one. It is global, it can be done anywhere and its results are valid everywhere. Most importantly, it is not the possession of any elite or special cadre.

If this book has succeeded, or failed—I get them mixed up—then it will have given you some new ways to think about science that respects your expertise as a member of the approximately 150th generation of recorded humanity. Hope you're ready to join in the fun.

New York, USA

Cambridge, UK

December 31, 2014

Notes and Works Consulted

I have avoided scattering footnotes through the text because I feel that it interrupts the reader's flow, often unnecessarily. I have instead included a set of notes and remarks here, keyed to page numbers that you can use if you find them interesting. My suggestion is to skim through the notes and if you find one that interests you go back to the page in the book that it refers to and read them together. In most cases sources that I refer to in the text are easily found online with a simple Google search, and so I have not given full citations. In cases where material came directly from a book that is now out of print or a paper that I could not find easily online, I have included the relevant material as a pdf file on my website:

http://stuartfirestein.com

If I have omitted critical material or a piece of information is obscure in some way, please email me and I can fix that on the website.

I have also included a list of books and papers that I consulted and that I think are worthwhile reads. My remarks on them should be considered personal opinions.

INTRODUCTION
Page 1

The epigraph, from Benjamin Franklin, arguably America's first scientist, is taken from his report to the King of France on Animal Magnetism in 1784. The longer version of it is:

> Perhaps the history of the errors of mankind, all things considered, is more valuable and interesting than that of their discoveries. Truth is uniform and narrow; it constantly exists, and does not seem to require so much an active energy, as a passive aptitude of the soul in order to encounter it. But error is endlessly diversified; it has no reality, but is the pure and simple creation of the mind that invents it. In this field the soul has room enough to expand herself, to display all her boundless faculties, and all her beautiful and interesting extravagancies and absurdities.

Page 4

Peter Medawar, Is the scientific paper fraudulent? *The Saturday Review*, August 1, 1964, pp. 42–43. (A pdf of the article is on my website.)

Medawar won a Nobel Prize in 1960 for his work showing how our immune system recognizes self from other, the basis of graft and tissue rejection. His work, all on the fundamental mechanisms of what is known as acquired immune tolerance, was nonetheless critical to the progress of organ and tissue transplantation. Medawar has been called the father of transplantation, a designation he always disavowed.

He was equally well known, particularly in the United Kingdom where he lived and worked, as a brilliant spokesperson for science with an ability to explain complex ideas in understandable, accessible, and, perhaps most importantly, entertaining ways. Richard Dawkins has called him the "the wittiest of all scientific writers." He was well known among the British public for his television and radio appearances as well and his numerous popular books, all still worth a read.

He was also a vocal champion of the ideas of the philosopher Karl Popper with whom he maintained close ties. Popper shows up later in this book.

CHAPTER 1: FAILING TO DEFINE FAILURE
Page 7

The Gertrude Stein quote is from her collection of stories *Four in America*, written in 1933, but not published until 1947, a year after her death.

Encyclopédie, ou dictionnaire raisonné des sciences, des arts et des métiers (*Encyclopaedia, or a Systematic Dictionary of the Sciences, Arts, and Crafts*) was originally published in 1751 with numerous updates and revisions through at least the 1830s. I am not in general in favor of encyclopedias as they tend to freeze knowledge, concentrating on what's known rather than on the far more interesting what's unknown. However, this is a classic of the Enlightenment and, more than a mere compilation, a real attempt to define and understand many new ideas.

Pages 12–13

Both forms of gravity are in fact due to acceleration in a straight line. The moon is constantly falling toward the Earth but in doing so is actually accelerating in a straight line in curved space and thus never falls into the Earth's surface which is curving away from it. Similarly, in an elevator you're falling in a straight line and are weightless—at least until you hit bottom. I am indebted to Huw (pronounced Hugh) Price for this explanation, and I can hope only that I have not butchered it beyond recognition to make the simple point I am making here. Professor Price, of Cambridge University, is a philosopher and historian of physics. I was fortunate enough to be able to sit in on his class at Cambridge while on sabbatical there in 2013–2014. We are all fortunate that he has generously posted his excellent lecture notes and many other materials on his personal website for you to peruse at any time free of charge (that is, at no cost; one must be careful about using the word *charge* in the presence of a physicist, especially a philosopher of physics). Huw has the ability to take very difficult concepts and make them very accessible. He also can think and communicate fascinating things about time, his real scholarly work. Most of his papers are available online (again at his website), and if you start reading them time will seem to zip on by as your brain is ever more challenged by truly extraordinary ideas. Thank you, Huw. And forgive me. (Although you can Google him, his website is simply prce.hu/)

Pages 14–15

There are numerous books on Haeckel, who has become something of a controversial figure. You can also find vast amounts of information about him and his ideas on the web. In addition to his immense scientific curiosity, he had remarkable aesthetic sensibilities. A testament of that is a gorgeous book of his prints compiled and beautifully printed in paperback format, *Art Forms in Nature: The Prints of Ernst Haeckel* (Munich and London: Prestel-Verlag Press, 2014). The book includes 100 color plates, with contributions by Olaf Breidbach and Irenaus Eibl-Eibesfeldt and a preface by Richard Hartmann. The famous embryo plate is curiously not included, but you can find that online.

CHAPTER 2: FAIL BETTER
Page 25

The opening quote is from one of Beckett's late short story collections, *Worstward Ho* (1983).

Page 26

The description of *Waiting for Godot* as "a mystery wrapped in an enigma" was, probably accidentally, taken from a Winston Churchill speech in which he characterized Russia as "a riddle wrapped in a mystery wrapped in an enigma."

CHAPTER 3: THE SCIENTIFIC BASIS OF FAILURE
Page 39

The opening quote is a lyric from *Wild Wild Things* by David Byrne and The Talking Heads.

Pages 43–44

"The Three Princes of Serendip" (sometimes spelled Sarendip) is a Persian fairy tale dating from 1302. In fact the three princes in the original versions were wise and thoughtful in that Sherlock Holmes sort of a way—using simple observation of minor details to surmise the nature of unseen things. The story shows up in Voltaire's *Zadig*, where it is used as an example of sagacity and the scientific method of observation and inference. Poe and Conan Doyle may also have been influenced by Voltaire's telling. However, the current (English) usage to mean simple good luck is due to Horace Walpole coining the term "serendipity" in a letter to his brother in 1754.

CHAPTER 4: THE UNREASONABLE SUCCESS OF FAILURE
Pages 49–50

E. P. Wigner, The unreasonable effectiveness of mathematics in the natural sciences. Richard Courant lecture in mathematical sciences delivered at New York University, May 11, 1959. *Communications on Pure and Applied Mathematics*, *13*, 1–14 (1960).

A Wikipedia entry on this article gives numerous responses and related papers that it has engendered. A pdf of the original paper is on my website.

Page 51

This essay appeared as a book chapter by James Robert Brown titled *Readings in the Philosophy of Science: From Positivism to Post Modernism*, edited by Theodore Schick, Jr. (London: Mayfield Publishing, 1999). This collection contains many classic essays on philosophy of science by many of the leaders in the field. But it's an expensive book. A copy of the table of contents for anyone interested can be found at

http://www.gbv.de/dms/goettingen/301131694.pdf

Page 99

Derek de Solla Price is somewhat of a hero of mine. I first ran into his out-of-print book *Little Science Big Science* (New York: Columbia University Press, 1963) while writing *Ignorance*. It was based on a series of lectures he had given in 1960. Almost 50 years later, they were still full of fresh ideas and relevant perceptions of the scientific literature. It is possible to buy this book and what was actually his first book, *Science Since Babylon* (New Haven, CT: Yale University Press, 1961), on the used-book market. More important, I have discovered that many of his papers and photographs have been collected and curated and are available for scholarship purposes through the Adler Planetarium in Chicago. Information is available at the planetarium website. I have also included a pdf file of the collection's contents on my website.

Then I found out that he was one of the first students to obtain a PhD in history and philosophy of science (1949) in the then newly formed department at Cambridge University—just where I was spending a sabbatical year while writing this book! There is a remarkable story in *Science Since Babylon* about Price as a graduate student. He was searching around for medieval manuscripts concerned with scientific instrumentation, particularly of the astronomical sort. At the Peterhouse Library, the oldest in Cambridge, there was a book on astrolabes that had been attributed to an obscure astronomer and elicited little attention. The manuscript had curiously been written in Middle English, not Latin, and, most important, was dated 1392. The significance of this was that Geoffrey Chaucer—yes, the medieval poet was also an astronomy buff—had published a well-known manuscript on Astrolabes in 1391. In Price's words, "To find another English instrument tract dated in the following year was like asking,

'What happened at Hastings in 1067?' The conclusion was inescapable that this text must have something to do with Chaucer." In fact, it turned out to be a handwritten manuscript by Chaucer following up on the earlier book. In fact, it is the only extant full-length manuscript in Chaucer's own handwriting—except for some fragmentary bills and papers, there is no other document in Chaucer's hand. Quite a find for a graduate student!

There are many other interesting stories and ideas in Price's books and they deserve to be reprinted—a project I hope to take on in the near future.

(http://www.adlerplanetarium.org/collections/)

CHAPTER 5: THE INTEGRITY OF FAILURE
Page 64

I am referring to a widely available piece by Robin Williams that can be easily accessed online in a YouTube video, (https://www.youtube.com/watch?v=pcnFbCCgTo4).

It may be one of his most hysterical stand-up routines, but before you show the kids, be warned that it is liberally laced with profanity. In some ways it is all the more poignant because of his suicide after years of battling depression. You can consider that a tragic failure, but it is not Mr. Williams's. It is a failure that we continue to work hard on in the neurosciences. But Mr. Williams triumphed for a very long time and was easily one of the funniest men to have ever walked the earth—suicidal depression and all.

CHAPTER 6: TEACHING FAILURE
Page 74

Ernst Mayr, *The Growth of Biological Thought: Diversity, Evolution and Inheritance* (Cambridge, MA: Harvard University Press, 1982). The quote referred to here can be found on page 20 of the paperback edition.

Several books on the subject of teaching science to nonscientists at the university level appeared in the 1930s and early 1940s as a result of a serious inquiry into it headed by James B. Conant when he was president of Harvard. Conant was a chemist by training and was deeply involved in many matters of national security and science, including the development of the atomic bomb. He was considered a transformative president of Harvard, serving from 1933 to 1940, after which he left to work for the government. Chief among his many educational interests was how science would fit into a classic liberal education. Several books were generated from committees and conferences that he convened. To my mind they are as relevant today as they were 60 years ago. I'm not sure that is a good thing, but it does tell us that the problems are not easily solved.

Among them are:

James B. Conant, *Science and Common Sense*. New Haven, CT: Yale University Press, 1951.

James B. Conant, *Modern Science and Modern Man*. New York: Columbia University Press, 1952.

I. Bernard Cohen and Fletcher G. Watson, eds., *General Education in Science*, with a foreword by J. B. Conant. Cambridge, MA: Harvard University Press, 1952.

James Bryant Conant, *Two Modes of Thought: My Encounters with Science and Education*. Credo Series. New York: Trident Press, 1964.

Page 75

Michael R. Matthews, Colin F. Gauld, and Arthur Stinner, eds., *The Pendulum: Scientific, Historical, Philosophical and Educational Perspectives*. Dordrecht, The Netherlands: Springer, 2005.

Partly reprinted from *Science & Education, 13*(4–5), and *13*(7–8).

Here's a sample of the Table of Contents:

Page 79

Hasok Chang, *Inventing Temperature: Measurement and Scientific Progress*. Oxford Studies in the Philosophy of Science. New York: Oxford University Press, 2004.

Page 88

The official citation of the article is:

Margaret Mead and Rhoda Métraux, Image of the scientist among high-school students: A pilot study. *Science*, *126*(3270), 384–390 (1957).

This article may be difficult to obtain without access to a university library system. I have placed a pdf file of it on my website.

CHAPTER 7: THE ARC OF FAILURE

I used numerous texts to trace the discovery of the circulation of the blood, and there are several accounts of Harvey's specific work in this area. I was guided through it all by the work of historian Charles Singer and especially his small but packed book, *A Short History of Anatomy and Physiology from the Greeks to Harvey* (New York: Dover Publications, 1957).

A more scholarly and though description of the development of anatomy and physiology from the ancients through Harvey can be found in the remarkable *Edge of Objectivity* by historian Charles Coulson Gillespie

(Princeton, NJ: Princeton University Press, 1960). This is perhaps the most comprehensive but readable history of Western science from Copernicus to modern quantum physics and biology. Coulson develops the history of the life sciences as embodied by Galen, Vesalius, and Harvey in the context of the parallel developments in physics from Copernicus through Galileo and Newton. He is also careful to point out all the misfiring's in the twisted road to understanding the circulation of the blood. His account is more detailed than anything I could present here, and for the interested reader I strongly suggest this source (you'll find this material in Chapter 11).

Page 109

I was fortunate to be in Italy in the summer of 2014 and near enough to make a detour to visit the anatomy theater in Padua. It has been preserved, or perhaps reconstructed, so that it is as it was when Vesalius dissected there. Calling the space a "theater" completely fails to capture the reality of the structure. It is a very steep, three-story wooden circular structure. The three "rows" are just wide enough to stand in, and there is a balcony rail just high enough to keep one from falling over, but low enough to lean over for a good look at the proceedings in the small circular area that held the dissecting table. Standing in the pit, one can easily imagine the stench that must have permeated the place. A near lethal combination of the relatively lax bathing habits of the era, the heat and sweat of 300 men crowded vertically into a space that would be smaller than the average bedroom, only higher, and the likelihood that at least some of the newer "students" would be puking from the stink of the corpse and the sight of the dissection. The dissection area itself was continuous with an area known as the kitchen, another unfortunate misnomer. The "kitchen" was where Vesalius and his assistants would prepare the corpse for the day's dissection. Once ready, and with the students in the gallery, the corpse could be wheeled into the pit at the vortex of the "theater" and the day's lessons could begin. I think all medical students today should be given a tour of Vesalius's famous operating theater—so that they would never ever complain about their own working conditions.

CHAPTER 8: THE SCIENTIFIC METHOD OF FAILURE
Page 119

I have noted elsewhere my fondness for quotes. But I have been surprised to find that their sources are notoriously difficult to ferret out. The one at the start of this chapter is typical. I was sure it was from Churchill, and it has often been ascribed to him. But there is absolutely no record, by him or someone listening to him, that he actually ever uttered this phrase—let alone made it up. Same for Lincoln, who often gets the credit. People like Churchill and Lincoln (and Voltaire and Edison) regularly seem to be the beneficiaries of these misattributions, perhaps because they were so charismatic and it often seems as if they should have said something like this or that. It's like Yogi Berra said, "I never said half the things I said." Or at least I think that was Yogi . . .

Pages 130–131

This notion of creativity arising from the *dissociation* of ideas rather then their association is not new. It was, to my knowledge, first stated at least suggestively by Wolfgang Kohler in his study of chimpanzee intelligence and problem solving. See *The Mentality of Apes*, trans. by Ella Winter (London: Kegan Paul, Trench, Trubner; New York: Harcourt, Brace & World, 1925). It may have been Kohler who was the first to use the phrase "Aha experience" to describe insightful behavior. Kohler and his experiments show up again in chapter 14 on pluralism.

The notion of creative dissociation can also be found in an intriguing book with many radical ideas, at least some which are very current today, by Jose Ortega y Gasset, *The Revolt of the Masses* (New York: W. W. Norton, 1932).

Pages 132–133

Nobel laureate François Jacob passed away in April 2013, and his obituary in the *New York Times* is worth a read on its own. He was one of the great popular writers in biology—among the ranks of Sir Peter Medawar, Lewis

Thomas, E. O. Wilson, and Konrad Lorenz. I cannot recommend his short and deceptively simple books highly enough. They are full of beautiful ideas like "night time science." Here are my favorites:

François Jacob, *The Possible & the Actual*. New York: Pantheon Books, 1982.

François Jacob, *The Statue Within: An Autobiography*, trans. Franklin Philip. New York: Basic Books, 1988.

François Jacob, *The Logic of Life*, trans. Betty E. Spillmann. Princeton, NJ: Princeton University Press, 1993.

François Jacob, *Of Flies, Mice and Men*, trans. Giselle Weiss. Cambridge, MA: Harvard University Press, 1998.

CHAPTER 10: NEGATIVE RESULTS
Page 145

The epigraph is from a 1947 lecture Turing gave to the London Mathematical Society.

Page 161

There is an entertaining and very accessible book about this controversy by Elliot S. Valnenstein called *The War of the Soups and the Sparks: The Discovery of Neurotransmitters and the Dispute over How Nerves Communicate* (New York: Columbia University Press, 2005).

CHAPTER 11: PHILOSOPHER OF FAILURE

Popper published an enormous number of papers. I have consulted a few of them here, but mainly I relied on the knowledge of my friends at the Department of the History and Philosophy of Science at Cambridge. Thankfully I was there on a sabbatical or else I would have been totally lost.

One of the volumes of his collected works, the one I mostly used, is *Conjectures and Refutations: The Growth of Scientific Knowledge* (New York: Routledge, 1963).

CHAPTER 12: FUNDING FAILURE
Page 177

The epigraph is from Joseph Heller's novel *Good as Gold* (1979).

Pages 199–200

The ideas referred to here are spelled out in several places but mostly in this book:

Donald Gillies, *How Should Research Be Organised?* London: College Publications, 2008.

Gillies examines what in the United Kingdom is known as the RAE, or Research Assessment Exercise. His analysis, however, is easily mapped on to the United States funding situation.

Danielle L. Herbert, Adrian G. Barnett, Philip Clarke, and Nicholas Graves, On the time spent preparing grant proposals: An observational study of Australian researchers. *BMJ Open* 2013;3:e002800 doi:10.1136/bmjopen-2013-002800.

Using a variety of measures, this team of Australian researcher/economists arrived at the estimate of some 550 years' worth of time spent in preparing grant proposals. Given the much larger number of American researchers, the amount of time could easily be double that among American scientists. There are several follow-up papers to this one that are of interest as well. All are available on the web.

While writing this chapter, support for increased science funding came from a very unlikely source—an op-ed piece in the *New York Times* by former Congressman Newt Gingrich. He calls for a renewed emphasis on funding our major research institutions, claiming credit of course for having engineered the doubling of the NIH budget in the late 1990s. It was the most emailed article in that day's *Times*. Aside from the welcome support for increasing research funding, Gingrich marshals a lot of numbers that are useful to know. The piece can be found at: http://www.nytimes.com/2015/04/22/opinion/double-the-nih-budget.html?_r=0

CHAPTER 13: PHARMA FAILURE

There is no end to the papers and books written on the pharmaceutical industry, both pro and con. For the ugly side of the business, the go-to person is Ben Goldacre, the British physician and academic and science writer. He writes a column that appears regularly in the *Guardian* called "Bad Science," and he has also written two very popular books that are virtual exposés. The subtitle of his book *Bad Pharma* should tell you what the book is about. He also has a very compelling TED talk.

Ben Goldacre, *Bad Science*. London: Fourth Estate, 2008.

Ben Goldacre, *Bad Pharma: How Drug Companies Mislead Doctors and Harm Patients*. New York: Faber and Faber, 2012.

Goldacre's charges are generally well documented and cannot be ignored. To me they are the record of a tragic situation in which well-intentioned people trying to cure diseases find themselves corrupted by the orthogonal demands of investors. I don't know what the solution is. I wish I did, because it's very important. I don't mean to avoid these issues, but this chapter is more about the dynamics of failure in pharmaceutical industry research. And Goldacre seems to be doing a very competent job with the seamier side.

The numbers I have used are well documented and can be found in numerous published analyses. Here are four papers that I found relatively accessible and which had extensive bibliographies for further reading:

Bernard Munos, Lessons from 60 years of pharmaceutical innovation. *Nature Reviews/Drug Discovery*, *8*, 959–968 (2009).

P. Tollman, Y. Morieux, J. K. Murphy, and U. Schulze, Identifying R&D outliers. *Nature Reviews/Drug Discovery*, *10*, 653–654 (2011).

J. W. Scannell, A. Blanckley, H. Boldon, and B. Warrington, Diagnosing the decline in pharmaceutical R&D efficiency. *Nature Reviews/Drug Discovery*, *11*, 191–200 (2012).

P. Honig and S.-M. Huang, Intelligent pharmaceuticals: Beyond the tipping point. *Clinical Pharmacology & Therapeutics*, *95*(5), 455–459 (2014).

CHAPTER 14: A PLURALITY OF FAILURES

Berlin's writings are easily available, and all are still in print to the best of my knowledge. I found John Gray's book *Isaiah Berlin* (Princeton, NJ: Princeton University Press, 1996) to be an especially clear and helpful commentary and analysis of Berlin's work. Especially his philosophical and historical ideas (as opposed to his literary criticisms).

Other texts and works I used in preparing this chapter were:

Larry Laudan, *Science and Relativism*. Chicago: University of Chicago Press, 1990.

Nicholas Rescher, *Pluralism: Against the Demand for Consensus*. Oxford: Oxford University Press, 1993.

John Dupré, *The Disorder of Things: Metaphysical Foundations of the Disunity of Science*. Boston, MA: Harvard University Press, 1995.

S. H. Kellert, H. E. Longino, and C. K. Waters, eds., *Scientific Pluralism*, Volume XIX in the Minnesota Studies in the Philosophy of Science. Minneapolis: University of Minnesota Press, 2006.

Hasok Chang, Is Water H20? Evidence, Realism and Pluralism. New York: Springer, 2012.

EPILOGUE TO CHAPTER 14
Page 238

Wolfgang Kohler and the insightful problem solving of chimpanzees is detailed in his book *The Mentality of Apes*, trans. Ella Winter (London: Routledge, 2005). Still available in numerous versions.

Pages 238–239

The classic paper by BF Skinner on developing superstitious behavior in the pigeon. For all his care in not using mental states as scientific objects of examination, Skinner here uses the term "superstition" in the title, rather than "superstitious behavior" or "superstitious-like behavior." Even with the quotation marks, it seems a little sloppy. The paper is also a single-author paper, which would indicate he actually did the experiments himself. This is a bit curious since at this time he was Chair of the Psychology Department at Indiana University and was about to leave for a position at Harvard. You would think there would have been a laboratory full of students to do the experiments. I have never come across anything in his writings that refers to these experiments, and they seem in some ways to have been undertaken very casually. The paper itself is notably deficient in the details of his methodology. Read it for yourself.

B. F. Skinner, 'Superstition' in the pigeon. *Journal of Experimental Psychology*, *38*, 168–172 (1948).

This paper will be difficult for many to retrieve directly from the journal site without the resources of a university library. It is reproduced here:

http://psychclassics.yorku.ca/Skinner/Pigeon/

. . . and I have placed a pdf file of the original on my website.

Page 243

The potato-washing behavior in monkeys has had a somewhat checkered history, being at one time co-opted by certain "New Age" authors and expanded into something known as the "hundredth monkey effect." This work has been debunked repeatedly and has assumed the status of an urban legend. However, the original work of Kawamura and Masao Kawai details a legitimate scientific observation. There is one paper in English:

S. Kawamura, The process of subculture propagation among Japanese macaques. *Primates*, *2*, 43–60 (1959).

There are at least three other papers in Japanese that are not to my knowledge translated. This work has been extensively reviewed and extended by Kawai Masao and can be found in a book available in its entirety online:

http://link.springer.com/chapter/10.1007%2F978-4-431-09423-4_24

The book is:

Tetsuro Matsuzawa, ed., *Primate Origins of Human Cognition and Behavior.* Berlin: Springer, 2001.

The relevant chapter is: "Sweet Potato Washing Revisited," by S. Hirata, K. Watanabe, and K. Masao (pp. 487–508).

CHAPTER 15: CODA
Page 247

The epigraph is a paraphrase from one of her poems that Dove herself used in a talk about her work.

Rita Dove, "The Fish in the Stone," from *Selected Poems.* New York: Pantheon Books, 1993.

(For the record, the actual line is, "The fish in the stone/knows to fail is/to do the living/a favor.")

WORKS CONSULTED

This is a very partial list of the books or essays that I found most influential while I was thinking about and writing this book. There was a lot of other material and if my choice of material for the book were different or if you had asked me last year, "I might think otherwise." Nonetheless here are some recommendations.

1. Feynman, Richard P. *The Meaning of It All: Thoughts of a Citizen Scientist.* New York: Basic Books, 1998.

Three, often rambling essays based on a series of lectures given by Feynman at the University of Washington in 1963. These were published posthumously in this book. The lectures were titled "The Uncertainty of Science," "The Uncertainty of Values," and "This Unscientific Age."

Feynman always discounted philosophers and historians of science as just so many birdwatchers, but he then he regularly wrote books containing the highest levels of philosophical thought and historical background in science—and made them accessible to the general reader.

2. Chang, Hasok, *Is Water H20? Evidence, Realism and Pluralism*. New York: Springer, 2012.

Don't let the title fool you. This is a remarkable book about how we know that something is what it is. Starting with something we all surely know to be true, Chang shows us that most of us have no direct evidence that water is indeed H20 or any clear idea of what that actually means. Chang opens a can of worms and revels in how they crawl around, slithering through every bit of knowledge you thought you were sure of and making a slimy mess. What fun.

3. Berlin, Isaiah, *The Hedgehog and the Fox*. London: George Weidenfield & Nicholson Ltd., 1953.

Very easy to find in numerous paperback editions.

4. Berlin, Isaiah, *The Hedgehog and the Fox: An Essay on Tolstoy's View of History*, 2nd edition, ed. Henry Hardy, foreword by Michael Ignatieff. Princeton, NJ: Princeton University Press, 2013.

The most recent edition, and possibly the best.

5. Collins, Harry, *Are We All Scientific Experts Now?* Cambridge, UK: Polity Press, 2014.

6. Collins, Harry, and Evans, Robert, *Rethinking Expertise*. Chicago: University of Chicago Press, 2007.

Harry Collins is a sociologist of science at Cardiff University in the United Kingdom, and one of the clearest writers and thinkers on the subject of expertise today. *Are We All Scientific Experts Now?* is a short book (shorter even than this one) but is a first-rate analysis of the reasons that scientific expertise does not get the respect and deference it once did. Not that it necessarily should.

7. Livio, Mario, *Brilliant Blunders: From Darwin to Einstein—Colossal Mistakes by Great Scientists That Changed Our Understanding of Life and the Universe.* New York: Simon and Schuster, 2014.

Science writer Livio uses five of history's greatest scientists and shows how they blundered badly in some areas—although typically not the ones that made them famous. It's a good lesson in humility and helps to illustrate the way science often works and to dispel the "Smooth Arc of Discovery" myth that I have pointed out infects our educational curriculum and distorts the public's view of science. By the way, it's a fun book to read and very well researched.

8. Rothstein, Dan, and Luz Santana, *Make Just One Change: Teach Students to Ask Their Own Questions.* Cambridge, MA: Harvard Education Press, 2011.

I only wish I had come across this book while I was writing *Ignorance.* Rothstein and Santana have put together a straightforward and accessible book about what seems like a simple idea—get kids to ask questions, questions they care about. Don't be deceived. It's one thing to get kids or anyone to ask a question or two; it's another to get them take possession of the questions, to recognize that learning is asking questions and not just memorizing stuff. The art of making questions, nearly lost, is thankfully revived in this book. And if it seems like this is aimed exclusively at young kids, I heard Rothstein give a brilliant lecture to the Harvard Medical School curriculum symposium. But it would be a shame to wait until medical school to get started on this educational strategy.

9. Schulz, Kathryn, *Being Wrong: Adventures in the Margin of Error.* New York: HarperCollins, 2010.

Among all the self-help type books about failure and error I found this one to be the most interesting (although I really didn't read any of them all the way through). Ms. Schulz thoroughly analyzes error, from the emotional responses that accompany them to the reality of false outcomes. She looks at error from a sociological, personal and professional perspective and she draws on many sources both literary and historical. She does not spend much time on science, which is perhaps why I found the book to be engaging.

Index